*HISTORY, FOUNDATIONS OF ANALYSIS, T(15-16: 1), S, B, L. *The Development of the Foundations of Mathematical Analysis from Euler to Riemann.* I. Grattan-Guinness. MIT Pr, 1970, 186 pp, $10. Exactly what the title indicates. An historical development of the foundations of the basic concepts in mathematical analysis: theories of functions and continuity, limits, convergence of series (including Fourier series), as well as the formation of the derivative and integral. The theme is developed around the solution of the vibrating string problem. The aim is "not to give a detailed account of the discussion itself, but to show how it was influenced by, and then influenced, the development of the foundations of analysis of its time." Recommended as resource material for a reading course or seminar for post-advanced calculus students. Also, teachers of analysis should take note. R.B.K.

THE DEVELOPMENT OF THE FOUNDATIONS OF MATHEMATICAL ANALYSIS FROM EULER TO RIEMANN

The MIT Press
Cambridge, Massachusetts, and London, England

I. Grattan-Guinness

THE DEVELOPMENT OF THE FOUNDATIONS OF MATHEMATICAL ANALYSIS FROM EULER TO RIEMANN

Set in Monotype Times Roman
Printed and bound in the United States of America by
The Colonial Press Inc., Clinton, Massachusetts

ISBN 0 262 07034 0 (hardcover)

Library of Congress catalog card number: 75-110228

This book is dedicated to my wife,
for her help and encouragement during its preparation

INTRODUCTION

Our account of the historical development of mathematical analysis concentrates only on its foundational aspects. We start in the eighteenth century with the achievements and difficulties of the calculus epitomized in the mathematical examination of the motion of the vibrating string, and from there we follow lines of development which led in the early decades of the nineteenth century to the broadening of the subject into what we now know as "mathematical analysis": theories of functions and continuity, limits, and the convergence of series (including Fourier series), as well as the formation and use of the derivative and the integral. The story is divided into six chapters, according both to the selection of problems and also to the historical circumstances of their progress, and an appendix is given over to the introduction of convergence tests. From time to time we meet other branches of mathematics which were related to or which applied the new ideas in analysis, but their detailed history has not been attempted here, on the grounds that they would constitute digressions too extensive from our main theme; instead, either a brief summary has been given in a footnote, or else reference has been made to other sources which do discuss the history of the problem in question. Of especial importance in this context is the fact that this study itself ends with the work of Riemann on the foundations of the calculus, for his other ideas and their consequences led to a new era in analysis (of which set theory and measure theory were the most important components), whose own development would require a separate volume. In addition, we have not attempted to recount the many-sided applications of the techniques in classical analysis itself.

One aim of this study is to complement the textbook treatment of classical analysis: it could be read by students with a knowledge only of solution by functional methods and separation of variables of linear partial differential equations, and of the general principles and development of classical analysis itself. The key to the presentation is *motivation*; and it is here that its complementarity to textbooks may be most evident. The mathematical teaching literature is often criticized for prejudice of presentation in favor of current conventions, or simply for being boring; and the reason for these deficiencies seems to be that, however skillfully the subject matter is presented and the examples are chosen, there is always a fatal lack of problem-background for it all. One may learn of this theorem and that property, but the problems which they are supposed to be solving are not mentioned. Consequently there is no background of ideas, no perspective against which the results can be cast, both in their own right and relative to each other. The current ways of thinking about possibly old problems and theories are expounded with perhaps meticulous accuracy; but competing approaches and their alleged inferiority are either not mentioned or, worse, dismissed as unreliable or inadequate without the reasons for this action being explained.

We attempt to fill these gaps, not only by describing the historical background involved but also by including from time to time an interpretative section on, say, a particular problem *as a problem* or the special virtues or vices of a theory just described. For example, as well as recounting in chapter 6 the advent of the regime of noninfinitesimal analysis which is still dominant, we discuss in chapter 3 the difficulties in manipulating infinitesimals in earlier periods with the technical equipment then available. Again, the appendix on convergence tests begins with a general discussion of the logic of necessary and sufficient conditions.

Closely associated with the idea of motivation is that of *different levels of understanding* of both the problems that have been motivated and the ideas that have been put forth to solve them. We do not describe a serene passage of success from one certain result to another on a foundation of impeccable security, for if we did our account would neither perform its task of complementarity nor describe the historical development with any degree of accuracy whatsoever. Problems and theories worth studying are difficult, and their understanding may well be hampered by doubtless unintended preconceptions or adoption of attitudes learned from completely different problems. Ideas passed by one writer to another may do so, on the one hand, with a loss of understanding of their significance; on the other hand, they may be transformed and extended into techniques of far greater power. There are examples of both kinds of process here, and we see emerging not so much a success story as a *guesswork story*, with hopes for and belief in the achievement of better guesses. Therefore we find on occasion personal elements also; for mathematics is a human activity as liable as any other to uncertainty, rivalry and personal bitterness, and also to generosity, to the sharing of ideas, and the acknowledgment of the genius in others. These factors have been included not only for historical interest, but also because in some cases they appear to have affected the content of the results under discussion (or at least the manner of their presentation) and therefore play a role of genuine significance in the story.

Personality problems belong specifically to the historical arena; but it must not be felt that the only source of motivation for mathematics is to be found in its history. Certainly the history is an important source, perhaps the most important in the sense that sooner or later it has to be brought into consideration of the development of a theory; but one can often use a heuristic rather than a historical approach. To take an important example from analysis itself, the historical origins of set theory lie largely in a problem of Fourier series belonging to the post-Riemannian era of the subject. In due course set theory moved into wider and wider areas of mathematics, and has now become the new dogma for teaching the subject at more and more elementary levels. This is not the place to criticize the practice, except to say that, as usual, motivation is almost entirely absent—that is to say, the special features of set theory which give it superiority over "older" methods are never mentioned in the (comparatively few) cases where superiority at these levels can be found. But if we accept that its enriched theory of relations gives it some role to play in teaching mathematics at a comparatively early stage, then a historical motivation is completely out of the question: the amount of analysis that would have to be understood first would be far too substantial and difficult for the student, and quite irrelevant to the aims of the teacher. Indeed, a "simplified" historical presentation would very likely turn out to be oversimplified. In general, the original investigations of a subject often involved notions and techniques—and even problems—which are now completely forgotten, and any valid historical account at any level must not fail to resurrect these lost ideas to a degree sufficient for their significance to be understood.

A few remarks would be useful on the organization of the text. Every theorem, definition, diagram, and equation has a first number to denote the chapter, and then a decimal number to indicate its position in the chapter: thus we have theorem 4.3, diagram 2.1, equation (5.77), and so on. In the appendix the numbering is prefixed by "A": there we have theorem A.11, etc. The mathematics is presented throughout in notations of current popularity: the original notations and symbolism were usually different, and are discussed only when they reflect some important advance (or retreat!) in thinking. This is done in footnotes, which are used to discuss points of minor or peripheral importance, problems connected with the sources used, or matters of controversy with other commentaries. But the main purpose of the footnotes is to give the source references for all the results and ideas mentioned in the text. A substantial proportion of the works involved have appeared several times over in one form or another, either in new editions by the original author or in anthologies of some kind, and especially in editions of the author's Collected Works. All but one of the sources cited have been examined, and usually in all their various appearances; but in order to minimize the listing of separate page references to all these locations, the passage in question has been located by its article number, or given by the original equation or theorem numbers, when this has given a sufficiently accurate specification.

The number of cited works has turned out to be sufficiently large to require that each one be identified in the footnotes by a year date, with the full details (including, where appropriate, the main reappearances) given in the bibliography. The date in question is usually the official date on the title page of the book or journal concerned, but where it differs from the date of composition to an extent sufficiently important to merit attention in the main text or footnotes, then the latter date has been used for reference purposes. But in the following cases it has not been possible to use a dating code at all and so the work has been denoted by a catchword:

1. for books and personal journals published in more than one edition and/or volume over a number of years. The number of the edition has been indicated by a subscript to the catchword and the relevant date given after the volume number, for example:

Lacroix *Treatise*$_2$, vol. 3 (1819), ...

2. for manuscripts either published long after composition or still unpublished, whose dates of composition have not been established accurately. Manuscripts whose dates are known have been denoted in the usual way.

In addition, volumes from editions of the Collected Works of an author are not given a year date, as in general the appearance of such editions did not have a direct bearing on the historical development concerned. Points of interest concerning some editions are discussed from time to time, however, and the general details of publication are given in the bibliography.

The history of mathematics is a subject which has always suffered severely from a shortage of manpower, particularly when the mathematics involved is close to contemporary teaching and research. Therefore this study has been fortunate in the encouragement received during its preparation, and especially in the detailed and penetrating criticisms of Sir Edward Collingwood and Dr. J. R. Ravetz.

June, 1969.

ISSUES IN EIGHTEENTH-CENTURY ANALYSIS—THE VIBRATING STRING PROBLEM

1

"Eighteenth-century analysis" is a term which covers many problems in mathematics. The problem of the motion of the vibrating string is not one of them, for it belongs to the realm of mathematical physics; but the vigorous and even impassioned discussion which was given to its solution during the second half of the century was to have a profound effect on the analysis that was used. Our aim, therefore, is not to give a detailed account of the discussion itself but to show how it was influenced by, and then influenced, the development of the foundations of the analysis of its time.

First, the problem. A uniform elastic string is stretched between two fixed points A and B a distance l apart and put into small horizontal vibration. We impose on the system a Cartesian axis system with AB as x-axis and a horizontal line through A perpendicular to AB as y-axis, and attempt firstly to find an equation to represent the motion in terms of y as a function of x and of time t, and then to solve it to find an explicit expression for y as a function of x and t.

The problem is a beautiful one: indeed, it is one of the most outstanding problems in the history of mathematical physics. The equation representing the motion,

$$\frac{\partial^2 y}{\partial x^2} = \frac{1}{c^2}\frac{\partial^2 y}{\partial t^2}, \tag{1.1}$$

discovered by Jean le Rond d'Alembert (1717–1783) and published by him in his opening paper of 1747,[1] was never called to question; but the generality of the competing solutions offered to it led to a lengthy controversy in which the whole of eighteenth-century analysis was brought under inspection: the theory of functions, the role of algebra, the real-line continuum and the convergence of series, as well as the physical interpretation of the solutions as string motion. So the affair was one of great importance; and because of its importance, it has been given much attention by historians of mathematics.[2] Our aim is to supplement their versions with an account written from the point of view described above.

We begin with mathematical techniques. Leonhard Euler (1707–1783) popularized the use of partial differentiation, and obtained in 1734 the results

[1] D'Alembert *1747a*, arts. 1–5. In his analysis $c=1$, and he uses the arc length variable s rather than x.

[2] The major accounts of the problem are, in chronological order: Riemann *1866a*, art. 1; Gibson *1892a*, pp. 137–146; Burkhardt *1908a*, pp. 1–50; Jourdain *1913a*, pp. 669–675; Truesdell *1960a*, pp. 237–300; and Ravetz *1961a*. We refer explicitly to these sources only in connection with individual points of interpretation and detail.

$$dz = \frac{\partial z}{\partial x} dx + \frac{\partial z}{\partial y} dy \tag{1.2}$$

and

$$\frac{\partial^2 z}{\partial x \, \partial y} = \frac{\partial^2 z}{\partial y \, \partial x} \tag{1.3}$$

for a function z of the variables x and y. Then he considered ways of integrating (1.2) for various forms of the coefficients of dx and dy, and so initiated the theory of solving partial differential equations.[3]

Integrating partial differential equations was the new step forward for mathematical physics. Euler was the leading pupil of John Bernoulli (1667–1748), who had been provoked by the Newton-Leibniz priority row over the invention of the calculus to devote the energies of himself and his school to the demonstration of the superiority of the Leibnizian system, especially in the description of the motion of continuous media according to Newton's law of universal gravitation. This was the post-Newtonian period of "rational mechanics": the investigation of the mechanics of rods, beams, fibres, membranes, fluids and many other such materials—including the vibrating string.[4] Euler's work on partial differentiation enriched the mathematical equipment needed for these problems; for since they usually involved at least one space variable as well as time, *partial*, rather than ordinary, differential equations were bound to be necessary.

D'Alembert (among others) found such equations for several physical problems. He also found an incoherent version of "d'Alembert's principle" which distinguished between internal and external forces on any element of the body and so made it clear that the internal forces cancelled each other out (by Newton's third law of motion), whereas the external forces gave rise to the resultant motion (as described by Newton's second law).[5] But of all these equations, the "wave equation" (1.1) was the only one he could solve. By a lengthy version of now familiar reasoning he found the functional solution

$$y = F(x + ct) + G(x - ct) \tag{1.4}$$

[3] Euler *1734b*. esp. art. 6. and the sequel *1734c*. See Cajori *1928a* for some account of the fragmentary study of partial differentiation and partial differential equations before Euler.

[4] Truesdell *1960a* gives an exciting account of the whole development of rational mechanics from Galileo (1638) to Lagrange (1788).

[5] See Truesdell *1960a*, pp. 159–160 on anticipations of the principle, and pp. 186–191 on the vagueness of d'Alembert's own understanding of it. Truesdell makes it clear that d'Alembert failed to exploit it (see esp. p. 238).

(where F and G are arbitrary functions). The boundary conditions at A and B:

$$y=0 \quad \text{when} \quad x=0 \quad \text{and} \quad x=l \quad \text{for all } t. \tag{1.5}$$

reduced the solution (1.4) to

$$y=F(ct+x)-F(ct-x), \tag{1.6}$$

where F is of period $2l$. If, in addition, motion began from the taut position, that is,

$$y=0 \quad \text{when} \quad t=0 \quad \text{for all } x, \tag{1.7}$$

then (1.6) would show that F was an even function.[6]

So we must give credit to d'Alembert for a brilliant analysis of the problem: the first genuine advance beyond the solution

$$y=K\sin\frac{\pi x}{l} \tag{1.8}$$

suggested by Brook Taylor (1685–1731) thirty years before for the steady motion situation.[7] But then he followed with a rambling sequel on the kinds of function which were allowable in the solutions. Only one point needs to be made here; but it is an important one, for it heralds the first challenge to eighteenth-century analysis that the vibrating string problem was to offer.

D'Alembert insisted that all functions used in the solution should be differentiable.[8] To understand why he made this stipulation, we must examine the theory of functions which was generally adopted at this time. The aim of Bernoulli and his followers was to demonstrate the superiority of the Leibnizian calculus over other methods of tackling physical problems. Now the calculus with which they worked was essentially a *calculus of operators on algebraic expressions*; the operators were those of differentiation and integration, whose inverse character had been the great discovery

[6] D'Alembert *1747a*, arts. 7–10. The account in Burkhardt *1908a*, pp. 10–11, seems to be an idealization of d'Alembert's argument.

[7] See Taylor *1713a* and *1713b*, and especially *1715a*, pp. 89–93. See also the equivalent derivation of the result by means of an n-body analysis in John Bernoulli *1727a* and *1728a*.

To say with Truesdell that "little but confusion" could follow from Taylor's difficulties seems to belittle both the progress that he did make on the problem, with little previous work to draw on, and also the implications of his solution (1.8) which were to take a central place in later developments (see Truesdell *1960a*, pp. 129–132).

[8] D'Alembert *1747b*: see esp. arts. 15–18 where he classified functions according to their convexity and concavity, that is, their different kinds of differentiability.

of both Newton and Leibniz. The two problems of tangent construction and area evaluation, which previously bore a relation to each other no closer than that of a similarity of type, were now twins, linked by an "inversion principle"; the powerful algebraic calculus allowed the mathematician to move easily along a whole chain of integrations and differentiations of a function according to his needs. But with power there is always responsibility; and in this case the limitation was that every operation must take place on a function which obeyed a "law of continuity" (that is, of differentiability). Thus the calculus was understood to operate validly only on those functions which fulfilled these conditions, and they were the differentiable functions: polynomials, trigonometric and exponential functions, and all such algebraic expressions which *yielded a definite result* from each operation of the calculus. This was the foundation of the theory of functions which d'Alembert and his contemporaries learned and created, and which they assumed to be true under the extension into several variables that was required by problems such as that of the vibrating string.[9]

Euler was moved by the beauty of the problem to publish his own version of it in the following year. His interest in it was so great that, most unusually for him, he published both Latin and French versions of his paper.[10] The solution he found was identical with d'Alembert's (and the derivation was superior); but then he took issue with d'Alembert over the specification of the arbitrary function F in (1.6), claiming that it could be deduced completely from knowledge of the initial position and velocity functions (and periodicity). To put the point in modern notation, if

$$y = h(x) \quad \text{and} \quad \frac{\partial y}{\partial t} = k(x) \quad \text{when} \quad t = 0, \tag{1.9}$$

then, from (1.6),

$$F(x) - F(-x) = h(x), \tag{1.10}$$

$$F(x) + F(-x) = \frac{1}{c}\int^{x} k(t)\, dt, \tag{1.11}$$

which equations specify F in terms of h and k, as required.[11]

[9] For example, Euler took it for granted that his equation (1.3) was *always* true; for z is a function "however composed" of x and y. (See Euler *1734a*, art. 6.)

[10] Euler *1749b* is the Latin version, *1748a* the French. The title page of the latter indicates that it is a translation of the former.

[11] Euler *1749b* (or *1748a*), arts. 11–29.

But how "general" are h and k? Euler now moved away from d'Alembert, for he allowed into the problem functions which did *not* necessarily have a derivative at every point—"functions with corners," as we may call them. The reason for this extension is clear from the physical situation: the initial position function h, especially, *must* include such functions, to cover the case of the motion being started by the plucking of the string. But in making this demand of the solution, Euler went for the first time beyond the scope of the theory of functions which he had done so much to consolidate and enrich. He was well aware that the extension held great promise for analysis, for later he wrote to d'Alembert that "considering such functions as are subject to no law of continuity opens to us a wholly new range of analysis."[12] For the new theory marked something of a *return to geometry*, a theory of shapes to replace a theory of algebraic expressions; but there were features of it which were to evade his proper understanding and so cost him the full discussion of the vibrating string problem.

Euler explained the situation in more detail in his textbook on analysis, which he published during 1748. The first volume was devoted to the algebraic theory of functions—where, among other things, he made standard the practice of denoting constants by a, b, c and variables by x, y, z, and introduced his classifications of functions as algebraic or transcendental, explicit or implicit, and uniform or multiform (that is, single- or many-valued).[13] But he began the second volume with the extension of this theory to which the vibrating string problem had led him, by making a distinction between a "continuous" and a "discontinuous," or "mixed," curve (or function).[14]

Euler's terminology is different from ours (which, as we shall see, has nineteenth-century origins), and the difference is important. Euler's term "continuous" is synonymous with our "differentiable," and refers to the old theory of functions; but his "discontinuous" corresponds to our "continuous," and includes the new functions *each of which was to be*

[12] Letter from Euler to d'Alembert, December 20th, 1763, quoted in Truesdell *1960a*, p. 276. The letter is summarized in d'Alembert *1768b*, art. 16, which is also in Euler *Works*, ser. 2, vol. 11, sect. 1, p. 2.

[13] Euler *1748b*, vol. 1; see especially arts. 1–4 and 8–18. For a list of the major innovations in the work, see Boyer *1956a*, p. 180.

[14] Euler *1748b*, vol. 2, art. 9. The motivation to this new distinction from the analysis of the vibrating string is mentioned in Euler *1763a*. (See especially the summary and art. 19; and also art. 3 for another explanation of the theory.)

Euler: "continuous" Euler: "discontinuous"

Modern: "differentiable" Modern: "continuous" Modern: "discontinuous"

DIAGRAM 1.1

considered as being composed of the algebraic expressions appropriate to their "continuous" portions. (See diagram 1.1.)[15]

D'Alembert was very critical of this extension: in a reply of 1750 he repeated his view that a solution was possible only "for the cases where the different shapes of the vibrating string can be included in one and the same equation. In all the other cases it seems to me to be impossible to give to y a general form."[16] And later he was to add to his statement of faith a genuine criticism of Euler's extended theory of functions: the difficulty of interpreting the wave equation itself at a "corner point." To put the criticism in modern terms, the problem is that the left- and right-hand derivatives of y with respect to x at such a point are different in value and consequently the second derivative $\frac{\partial^2 y}{\partial x^2}$ is not defined there. Consequently how can it be said to equal $\frac{1}{c^2}\frac{\partial^2 y}{\partial t^2}$?[17]

Euler's reply displayed another point of difference with d'Alembert: his treatment of the real-line continuum. Euler was an infinitesimalist: he used infinitesimally small quantities in his mathematics, even formulating laws such as

$$a + dx = a, \qquad \sqrt{dx} + dx = \sqrt{dx}, \qquad dx + (dx)^2 = dx \tag{1.12}$$

[15] Euler does not appear to have considered functions which are discontinuous in our sense at all: indeed, there would seem to have been little need for him to do so in his problems.

[16] D'Alembert *1750a*, art. 2. See also art. 1, where he presents the functional solution afresh by separating the variables of the wave equation. This seems to be the first use of this method in the solution of partial differential equations (see Truesdell *1960a*, p. 241).

[17] D'Alembert *1761a*, art. 10; the point is made in an obscure way.

as part of his general considerations.[18] But he missed the significance of d'Alembert's objection by invoking infinitesimals to dismiss it. "I do not deny" he told d'Alembert, "that in applying the calculus one commits some error; but I maintain that in totality this error becomes infinitely small and entirely nothing,"[19] and in a later letter he applied this principle to argue that since the vibrations were small anyway, the corner angles would be small and so the curve would differ only infinitesimally from a "continuous" one; but d'Alembert, suspicious of any use of infinitesimals in analysis, was not convinced.[20]

So we end the first part of the controversy and enter upon the second, which also has its origins in Euler's 1747 reply to d'Alembert. In looking for functions for his solution which the periodicity condition satisfied, he turned naturally to Taylor's sine solution (1.8) and considered the more general

$$\alpha \sin \frac{\pi x}{l} + \beta \sin \frac{2\pi x}{l} + \cdots. \tag{1.13}$$

Whether he thought the series was finite or infinite is completely unclear; but he was sure that it held only for "continuous" (that is, differentiable) functions, and so could not compete with his own solution.[21] However, it was considered much more seriously in 1753 by a new contestant: John Bernoulli's son, Daniel Bernoulli (1700–1782).

Daniel Bernoulli had been impressed by Taylor's identification of his solution (1.8) with the frequency of vibration of the string, and he offered a physical argument in favor of trigonometric series based on a prin-

[18] See Euler *1755a*, part 1, ch. 3, esp. arts. 83–97. For thoughtful criticisms of Euler's treatment of infinities and infinitesimals see l'Huilier *1786a*, esp. ch. 11, a sincere but seemingly uninfluential essay on the state of the calculus at that time.
[19] Euler *1765a*, art. 6.
[20] The letter was quoted by d'Alembert in his *1768a*, art. 24, which is also in Euler *Works*, ser. 2, vol. 11, sect. 1, p. 2. The solution to this problem of interpreting the wave equation as an *integral* equation only came a century later from E. Christoffel (see Truesdell *1960a*, pp. 290–291). For d'Alembert's views on infinitesimals, see *1767a*, especially pp. 249–250. For his general views on analysis, see the articles "Différentiel," "Fonction," and "Limite" in the *Encyclopédie ou dictionnaire raisonné des sciences, des arts et des métiers* which he edited with Diderot. In "Cordes (vibration des)" he preached his own interpretation of the functional solution of the wave equation against Euler's. (D'Alembert's articles are all signed "(O)" in the *Encyclopédie.* On the entire work, see Grimsley *1963a*, pp. 1–77; and the articles in the *Revue d'Histoire des Sciences*, vol. 4 (1951), pp. 201–378.) Versions of these articles also appeared in the 1780s in the three volumes of the *Encyclopédie méthodique: mathématiques*, of which he was an editor.
[21] Euler *1749b* (or *1748a*), art. 30.

ciple of superposition of frequencies of vibration. According to him, the string "can form its uniform vibrations in an infinity of ways" with a multiplicity which "tends to infinity," implying that "all sonorous bodies contain potentially an infinity of sounds and an infinity of corresponding ways of making their regular vibrations." So the string can superimpose all its Taylorian harmonies in all its possible combinations, implying that the trigonometric solution

$$y = \alpha \sin \frac{\pi x}{l} \cos \frac{\pi c t}{l} + \beta \sin \frac{2\pi x}{l} \cos \frac{2\pi c t}{l} + \cdots \tag{1.14}$$

(where the series is definitely infinite this time) gives the most general description of the motion. The initial position function

$$y = f(x) \quad \text{when} \quad t = 0 \quad \text{for all } x, \tag{1.15}$$

inserted into (1.14), yields Euler's equation (1.13) (when taken as an infinite series), and so Bernoulli's physical argument repudiates Euler's rejection of the trigonometric form.[22]

Bernoulli did not support his opinions with any mathematical arguments: thus there was no proof of the generality of the function capable of such representation, nor any indication of methods of calculating the coefficients. Euler published a quick reply to Bernoulli's assertions: this reply contained a crucial mistake which was to put the whole discussion off course. "But perhaps," he wrote, "one will reply that the equation [(1.13)], because of the infinity of undetermined coefficients, is so general that it includes all possible curves; and it must be acknowledged that if that were true, the method of Mr. Bernoulli will furnish a complete solution. But, apart from the fact that that great geometer has not made that objection, all the curves contained in that equation, although one increases the number of terms to infinity, have certain characteristics which distinguish them from all other curves."[23] Periodicity was one restricting property and, in the particular case of the sine series in (1.13), oddness was another.[24] Later he even urged the specific case of the function which is zero over part of the interval (corresponding to an initial

[22] Bernoulli D. *1753a*; see esp. arts. 1–4, 12–18 and 21–27. See the sequel *1753b* and also *1758b* for further discussion of the superposition of frequencies; and the letter to Euler, apparently written during the mid-1750s (see Truesdell *1960a*, p. 257), where he criticized mathematical reasoning in general, in Fuss *1843a*, vol. 2, pp. 663–665.
[23] Euler *1753a*, art. 9.
[24] Euler *1753a*, art. 9.

stimulation only over the rest of the string) as incapable of trigonometric expression because of the nonsinusoidal nature of that part.[25]

Euler had compared the merits of two arguments:

1. The "interpolation argument," as we may call it, which asserts that the infinity of undetermined coefficients in (1.13) could be chosen so as to fit the series to a "general" (including, presumably, a "discontinuous") curve. If this argument were correct, then indeed, (1.13) could lead to a general solution of the wave equation. *But* there is

2. The "periodicity argument," which Euler supported, which points out that the trigonometric character of the series places unavoidable and restrictive conditions on the representable function $f(x)$.

Euler's preference for the periodicity argument is seen to be completely mistaken when we remember that the analysis has reference only to the portion AB of the x-axis over which the string is stretched: what happens outside AB is irrelevant to the vibrating string, and therefore to the mathematics used to describe it. Hence the trigonometric functions are perfectly adequate to handle the problem (from this point of view, at any rate), since in (1.13) they are of period $l = AB$. Similarly the oddness of the sine terms is irrelevant to the problem, since that applies only to the section $(-\infty, 0)$ of the x-axis and not to AB at all.

So Euler had made a mistake; but it is not a trivial slip. His error is a great one, of the kind that only a deep thinker could advance sufficiently into a problem to make. It was at the heart of his new theory of functions, and showed the influence that the old theory still had upon him.

One of the basic features of the old theory was that the algebraic expressions involved were understood to operate over the *whole* of their range of definability. To take explicit examples: the function

$$y = +\sqrt{1-x^2}, \tag{1.16}$$

representing the upper half of the unit circle, was accepted as definable only when $-1 \leq x \leq +1$, since outside that range y took imaginary values. But a function such as, say,

$$y = x^3 \tag{1.17}$$

took real values over the whole of the x-axis, and so was to be thought of that totality. Now when Euler introduced his theory of "discontinuous" functions defined by means of their "continuous" segments, he never

[25] Euler *1765a*, art. 10; *1765b*, art. 22.

properly realized that to each expression which gave the "continuous" segment he would have to take the revolutionary step of assigning an interval of definability corresponding to the segment, *and completely independent of the algebraic form.* This was what the "return to geometry" of his new theory meant; but he never broke sufficiently free of his algebraic background to realize that the return was needed. The new theory did indeed open the "entirely new course" of which he had written to d'Alembert; but he never realized the direction the course would take.

These are the algebraic constraints which led him to his mistaken criticism of trigonometric series. For when he explained the grounds of the criticism in terms of their periodicity, he meant the *algebraic periodicity* of the sine terms over the whole x-axis instead of the *geometric periodicity* of $f(x)$, which is based on its behavior over $[0,l]$. Let us explain this distinction more fully. $f(x)$ is defined by the shape it takes over $[0,l]$ whether or not there is an algebraic expression to specify it over that interval; and if there is need to consider $f(x)$ outside $[0,l]$ (for example, in using $f(x \pm ct)$ to compare the corresponding series with the functional solution (1.4)), then the function is defined there by geometric periodicity of the original shape. But algebraic periodicity is the periodicity of the trigonometric functions relative to the whole line, irrelevant to the physical phenomenon taking place over AB: an algebraic solution to a geometrical problem, in fact, which it could not possibly solve.

The belief in algebra pervades the whole of eighteenth-century analysis. One sees it in the whole confusion over functions themselves. The distinctions between different kinds of function which matter mathematically are those of our diagram 1.1: the classes of differentiable, continuous and discontinuous functions hinge on the existence, and equality or non-equality, of the left- and right-hand limiting values of a function and its derivative at a given point. But such distinctions are essentially geometric or analytical, out of reach even of Euler, although his theory of "discontinuous" functions had them within its grasp. Instead one finds in eighteenth-century mathematical writings a profligacy of terms whose vagueness or even irrelevance to the problem of the "generality" of functions greatly obscured the results. "Algebraic," "transcendental," "obeying a continuous law," "mechanical," "pieced together," "drawn by hand," "regular," "totally discontinuous": so it went on, with the same term used for different classes of function, different terms used to denote the same class, and subdistinctions introduced which were useful in other contexts but basically without impact on the problem of general solutions to partial differential equations.

The situation is reflected faithfully in the discussion of the vibrating string. For example, the functional solution of d'Alembert and Euler suffered somewhat in understanding from the wealth of vernacular available, and especially from the additional trap of periodicity in its irrelevant algebraic repetition along the x-axis.[26] Again, Euler first examined trigonometric series in 1747 because he was looking for (geometrically) periodic functions in his solution, yet he rejected them in 1753 because of their (algebraic) periodicity. Meanwhile, Daniel Bernoulli, perhaps under the influence of Taylor's solution (1.8), suggested only series of sine terms because these were the only trigonometric functions which themselves passed through A and B: hence even the possibility of adjoining a cosine series with suitably chosen coefficients was not allowed for. Above all, periodicity meant an immediate dismissal of Bernoulli's trigonometric solution, so that technical matters such as its convergence or the calculation of its coefficients received no attention in their proper context.

Euler's reply to Bernoulli epitomizes the whole failure. The periodicity criticism caused him to ignore the genuine attack that can be made on Bernoulli's full solution (1.14): that it implies

$$\frac{\partial y}{\partial t}=0 \quad \text{when} \quad t=0 \quad \text{for all } x, \tag{1.18}$$

which is definitely less general than his own solution with its initial velocity function k in (1.11).[27] Instead he affirmed his algebraic faith as never before. He introduced the notation $\phi(x, t)$ for "a certain function of the abscissa x and time t," replaced the then-standard notation p, q, r, s, t, for first- and second-order partial differentials by the symbolic form "$\left(\frac{dz}{dx}\right)$" which reflected the process of differentiation that had taken place,

[26] This point is well exemplified by some of the diagrams in their papers. (See esp. d'Alembert *1747a*, diags. 1, 2, 9 and 12; *1750a*, diags. 4 and 5; and Euler *1749b* (or *1748a*), diags. 2 and 3; *1753a*, diag. 3.)

[27] The criticism can be met by incorporating sine t-terms as well as cosine terms into (1.14). In his historical account of Bernoulli's work, Riemann does make this incorporation in the solution to give

$$y=\sum_{r=1}^{\infty} \alpha_r \sin \frac{r\pi x}{l} \cos \frac{r\pi c}{l}(t-\beta_r); \tag{1}$$

but this would appear to be historical generosity on his part, especially as Bernoulli himself never wrote down the full solution (1.14). (See Riemann *1866a*, art. 1.)

and, above all, he concluded a complete reworking of the problem with the words: "see then to what the problem of the motion of the string is reduced: it is a question of finding for y such a function of the two variables x and t which satisfies [the wave equation] . . . we seek in general *all the possible values* of x and t which, being put for y, yield [that equation]."[28]

This was the intellectual atmosphere of the discussion: algebraic methods to solve nonalgebraic problems. Therefore it was led to confusion by its misunderstandings, among which the periodicity problem was dominant.[29] Yet all wás not yet over, for a new competitor was to enter the scene a few years later: a young Italian called Luigi de la Grange Tournier (1736–1813). Lagrange was born in Turin, but his family was of French origin and he used the Italian form of his name only in his first publication—a letter of his eighteenth year on successive differentiation and integration.[30]

Lagrange belonged to the new generation, and a long paper on the nature and propagation of sound in 1759 helped to make his early reputation; both d'Alembert and Euler, for their own reasons, urged that Lagrange should succeed Euler as Director of the Berlin Academy.[31] But the mathematical atmosphere that he inherited was to dominate his thought on the analysis of the vibrating string, for he accepted all Euler's views on it in favor of the alternatives. Thus he felt that Euler's extended interpretation of the functional solution to "discontinuous" functions

[28] Euler *1753a*, arts: 15, 17 and 22: italics inserted. Euler popularized the notation form $\left(\frac{dz}{dx}\right)$ in his later work, notwithstanding the ambiguity of the parentheses in expressions such as $\left(\frac{dz}{dx}\right)^2$; the modern $\frac{\partial z}{\partial x}, \ldots$ only came into general use toward the end of the nineteenth century. (For a review of many of the notations used for partial differentials at one time or another, see Cajori *1929a*, paras. 593–619.)

[29] Realization of the importance of the periodicity confusion is the principal lacuna in the historical accounts of the vibrating string problem. The only mention of it at all seems to be by Truesdell, where it is dismissed as a "wholly fallacious argument," which shows that Euler had "no idea that what we now call a nonanalytic function may be represented in a finite range by a trigonometric series" (Truesdell *1960a*, p. 261). True enough; but Euler's lack of ideas is a fundamental one, a real mistake of understanding rather than a passing aberration. The failure of the discussion is usually attributed to the confusion over generality of functions and the inability to calculate the coefficients of the trigonometric series. These are certainly contributory factors; but they are subsidiary to the periodicity question, since they come into prominence only when periodicity has been settled.

[30] Lagrange *1754a*.

[31] See Delambre *1812a*, pp. xlvii–xlviii; also in Lagrange *Works*, vol. 1, pp. xxi–xxii. See also the letters from d'Alembert to the King of Prussia in d'Alembert *Works*, vol. 5 (1822), pp. 260–267.

based on initial configuration and velocity functions was valid, and he followed Euler's criticism of Bernoulli's trigonometric series on the grounds of periodicity.[32] But he took issue with Euler's manner of deriving his result, and for a very interesting reason. "Discontinuous" functions were needed, without doubt; but Euler's method of obtaining them directly from the principles of the calculus was illegitimate.[33]

So Lagrange was dubious about limits and infinitesimals; and he took his doubts to the length of suggesting new foundations for the calculus of which he told Euler that "I believe to have developed the true metaphysics of their principles, as far as it is possible," in a letter of 1759 also informing him of his paper on the vibrating string.[34] The new approach was based on a theorem with which Brook Taylor is fortunate to be identified: the Taylor series

$$f(x+h)=f(x)+hf'(x)+\frac{1}{2!}h^2f''(x)+\cdots.^{35} \tag{1.19}$$

Lagrange's idea was to *define* the derivatives of a function as the coefficients $a_1, a_2, \ldots$ in the expansion

$$f(x+h)=a_0+ha_1+\frac{1}{2!}h^2a_2+\cdots, \tag{1.20}$$

a method which seemed to him to be "the most clear and the most simple that has yet been given: it is, as one can see, independent of all metaphysics and of all theories of infinitely small or vanishing quantities."[36]

How Lagrange thought his method would avoid "metaphysics" (that is, limits) or infinitesimals is difficult to say: after all, the coefficients have to be calculated somehow or other, and he made the assumption that a

32 Lagrange *1759a*, arts. 12–18. In general Lagrange tended to follow Euler in his mathematical style. Temperamentally, however, he was much closer to d'Alembert, with whom he maintained an active correspondence (published in Lagrange *Works*, vol. 13) until the latter's death in 1783.

33 Lagrange *1759a*, art. 15.

34 Letter from Lagrange to Euler, November 24th, 1759 (see Lagrange *Works*, vol. 14, p. 173). P. E. B. Jourdain thought that this passage referred to the validity of using infinitesimals (see Jourdain *1913b*).

35 There are precedents for Taylor's series in Newton, Leibniz and John Bernoulli. Taylor's version appeared in 1715 (Taylor *1715a*, pp. 21–23), although he knew it in 1712 (see Bateman *1907a*, p. 371); at all events, John Bernoulli was very annoyed about it (see Fleckenstein *1946a*). For an (incomplete) study of the development of Taylor's series, see Pringsheim *1902a*.

36 Lagrange *1772a*, art. 3. This paper was his principal work on his theory, but he first published his ideas in a footnote to a paper by P. Gerdil in 1762; that is, shortly after his paper on the vibrating string (Lagrange *1762a*).

function can be expanded in a Taylor series in the first place. (In fact, his analytical writings give the impression that he used his principle as an *excuse* for drawing on infinitesimals when convenient!) Nevertheless, he held this view till the end of his life.[37] Whether or not he had come to it at the time of writing his paper on sound is not certain; but as we saw, he felt sufficiently dissatisfied then with Euler's reasoning to urge a new approach to the vibrating string problem.

The fresh attack was based on analyzing n equal and equally spaced bodies linked by weightless cord, which would become the heavy string when n tended to infinity. Lagrange claimed a measure of originality of thought for himself in using this model, but in fact it was a fairly popular device of rational mechanics and had already been tried on the vibrating string by John Bernoulli and Euler.[38] Its advantage to Lagrange lay in the (partial!) avoidance of limit-taking: for example, instead of the wave equation, he started from the sequence of differential-difference equations

$$\frac{d^2y_k}{dt^2}=c^2(y_{k+1}-2y_k+y_{k-1}), \qquad k=1, 2, \ldots, n. \tag{1.21}$$

A full discussion of his long analysis[39] would form too extensive a digression from our theme. But there is one particular stage of the reasoning which is of great importance to us. Having obtained an explicit expression for y in terms of the initial positions and velocities of each body in the Eulerian manner, he took the number n to infinity to obtain the solution for the string itself:

$$\begin{aligned} y=&\frac{2}{l}\int_0^l \sum_{r=1}^{\infty} \sin\frac{r\pi X}{l}\sin\frac{r\pi x}{l}\cos\frac{r\pi ct}{l}\,Y(X)\,dX \\ &+\frac{2}{\pi c}\int_0^l \sum_{r=1}^{\infty} \frac{1}{r}\sin\frac{r\pi X}{l}\sin\frac{r\pi x}{l}\sin\frac{r\pi ct}{l}\,V(X)\,dX, \end{aligned} \tag{1.22}$$

[37] Apart from Lagrange *1762a* and *1772a*, see also *Functions*$_1$ (1797), arts. 1–3 and ff.; *1799a*; *Lessons* (1801), lessons 1 and 2; and *Functions*$_2$ (1813), part 1, arts. 3–9 and ff. Among his few supporters were Crelle, who tended to base all aspects of analysis on series (see esp. Crelle *1813a*) and prepared translations into German of Lagrange's textbooks on the subject (see Lagrange *1823a*, vols. 1 and 2), and Arbogast. (See Arbogast *1789a*. This manuscript seems never to have been published: Lagrange referred to it in the introduction to both editions of his *Functions*.) For an account of the method as compared with Euler's infinitesimals, see Yushkevich *1959a*; and on its generally critical reception, see Dickstein *1899a*. See also Boyer *1949a*, pp. 261–266.
[38] Lagrange *1759a*, introduction. See Bernoulli J. *1727a* and *1728a* (referred to in n. 7, p. 4); and Euler *1747a*, esp. arts. 31–44.
[39] Lagrange *1759a*, arts. 19–39.

where $Y(X)$ and $V(X)$ are its initial position and velocity functions.[40]

Equation (1.22) has caused more puzzlement to the historians of mathematics than most equations, for if we put into it the boundary condition

$$y = Y(x) \quad \text{when} \quad t = 0, \qquad 0 \leq x \leq l, \tag{1.23}$$

we obtain the "Fourier sine series" for the representation of the function $Y(x)$:

$$Y(x) = \frac{2}{l} \int_0^l \sum_{r=1}^{\infty} \sin \frac{r\pi X}{l} \sin \frac{r\pi x}{l} \, Y(X) \, dX. \tag{1.24}$$

But Joseph Fourier (1768–1830) made his discovery *half a century* after Lagrange's paper. Why did Lagrange miss it here?

To look for *one* answer to the question would be rash; but there would seem to be several contributory factors which together resolve the difficulty.

1. Equation (1.24) is a theorem of mathematical analysis applying to the physical model only at a particular moment in time; but Lagrange was not interested here in theorems of analysis, but in general solutions to represent physical motion over all moments of time. Therefore (1.24) would contain nothing of special interest to him.

2. In any case, (1.24) is *not* the Fourier sine series: the sum and integral signs need interchanging to give the correct equation:

$$Y(x) = \frac{2}{l} \sum_{r=1}^{\infty} \int_0^l Y(X) \sin \frac{r\pi X}{l} \, dX \sin \frac{r\pi x}{l}. \tag{1.25}$$

Lagrange would have had no qualms at making the change had he wished to; but in fact he had made the reverse interchange in the course of his deduction in order to obtain the "$\int \Sigma$" form of (1.22). In other words, Lagrange was not interested in an *infinite series* representation $\sum_{r=1}^{\infty}(\int_0^l \ldots)$ but in an *integral* representation $\int_0^l(\sum_{r=1}^{\infty} \cdots)$, an aspect of his apparent visualization of the vibrating string problem as a problem of *pulse propagation.* Another aspect of this view—and also an example of his confused thinking—is his version of our equation (1.22), with "dx" written for "dX." We have not followed Lagrange, since not only does "dx" make (1.22) incomprehensible but is in flat contradiction of the remarks he makes immediately following it: "where it is to be mentioned that the integrations must be done while supposing X, Y and V variables, and t

[40] Lagrange *1759a*, art. 37.

and x constants." Lest the point be thought merely a slip of the pen or the printer, we must mention that he used "dx" consistently in the latter sections of his analysis until he obtained the functional solution, and thus seems to have been thinking in pulse terms for his X, Y and V at a given point x and time t, in accordance with the above quotation. But the whole situation is muddled: the "dx" itself, for example, is finally eliminated by a sleight of hand with infinitesimals—by the critic of the foundations of Euler's calculus![41]

3. Equation (1.22), and hence (1.24), is anything but Lagrange's aim, for it represents a Bernoullian series solution with coefficients represented by Eulerian initial position and velocity functions. Lagrange was seduced along with Euler by the confusion over periodicity into rejecting series solutions outright: equation (1.22) was only a stage on the route to the functional solution which he wanted, and was no more important than any other stage. Having reached it, he made no pause for reflection on its potential uses, but pressed straight on, forward to the Eulerian goal.

So we see several causes of Lagrange's failure to spot the Fourier series, each of them indicating a different reason why he was not looking for it. The historians' puzzle is one they have created for themselves with the hindsight of Fourier's work.

There was still much to be said by the various contestants; but the standard of the discussion fell markedly, with entrenchment of position being the order of the day (though Lagrange did tend to move away from Euler's version of the functional solution toward d'Alembert's).[42] There were too many important points, especially periodicity, which were too imperfectly understood for more genuine advances to be made. Of the ideas on functions and general solutions to partial differential equations put forward by new men, the distinction between continuous and discontinuous functions (in our sense) by Louis Arbogast (1759–1803) was

[41] See Lagrange *1759a*, arts. 36–38: the pulse-propagation origin of Lagrange's thinking was urged in Ravetz *1961a*, pp. 84–85.

Commentators have found this proof of Lagrange to be full of "power," "beauty," and "virtuosity"—but at least this reader feels impressed by the fact that Lagrange knew what answer he was looking for before he started, and finds that he got it by creating a superficially sophisticated line of reasoning which actually develops with considerable shakiness, laced as it is with fortunate approximations and simplifications and a continuous vacillation over the changing statuses of its quantities as constants and variables which the case of "dX" epitomizes to perfection.

[42] The latter period is given detailed treatment in Truesdell *1960a*, pp. 273–300. In 1788 Lagrange published the first edition of his *Mechanique analitique*, allegedly the Bible of rational mechanics. The vibrating string problem was tucked away at the end of a section on n-body analyses. (See Lagrange *Mechanics*$_1$ (1788), pp. 333–337.)

the most useful; and the circumstances of Arbogast's work reflect well the situation at the time. In 1787, when only Lagrange of the contestants over the vibrating string was still alive, the St. Petersburg Academy (of which Euler had been Director until his death in 1783) proposed a prize problem to try to resolve the question of general solutions to partial differential equations:

"Whether the arbitrary functions which are achieved by the integration of equations with three or several variables represent any curves or surfaces whatsoever, either algebraic or transcendental, either mechanical, discontinuous, or produced by a voluntary movement of the hand; or whether these functions include only curves represented by an algebraic or transcendental equation."

Arbogast won the prize, for making a distinction between "discontinuity" and "discontiguity" of a function. The terminology could hardly have been worse; but the idea was a move in the right direction. His use of the terms "continuous" and "discontinuous" was Eulerian, but the new category of "discontiguous" was *our* use of the term "discontinuous." Thus he did go beyond Euler in considering "modern" discontinuous functions and supported—without genuine mathematical argument, however—Euler's advocacy of the validity of "discontinuous" (in his sense) functions in the solutions of partial differential equations.[43]

Despite this achievement the other issues were hardly touched upon, and thus when the Fourier coefficients for trigonometric series made their appearance during the century, they were never allowed to have any real

[43] See Arbogast *1791a*, esp. pp. 9–12 and 59–89. For commentary, see Jourdain *1913a*, pp. 675–678; and for information on Arbogast's life, Frechet *1940a*.

A few notes on the support for discontinuous solutions to partial differential equations on the latter half of the century may be useful. Predictably, Lagrange used functional solutions and exact differentials as developed by Euler (see Lagrange *1772b* and *1774a*; and compare with Euler *Integration*$_1$, vol. 3 (1770), parts 1 and 2). Laplace relied on discrete analysis, finding discontinuous solutions to difference equations and then taking the limit (see Laplace *1773a*; and *1779a*, arts. 18–23). By contrast Monge drew on geometrical arguments, interpreting the solutions as surfaces. (The genesis of this idea may be found in the manuscript published in Taton *1950a*, and in the letters from Monge to Condorcet in Taton *1947b*. On Monge's published work during the 1770s and 1780s, see Taton *1951a*, pp. 273–300.) Monge hoped to become the fifth competitor in the vibrating string problem during the 1770s, but d'Alembert refused to be drawn into a discussion with him (see Taton *1947b*, pp. 988–989). He was impressed by the functional solution, and built a physical model of it when teaching at the Ecole Polytechnique in Paris in the 1790s (see Monge *1809a*, p. 145). In *1787a*, Legendre criticized Monge's reliance on geometry as well as Laplace's analytical reasoning, which he replaced by an inversion of the technique in treating x, y, and z as functions of $\frac{\partial z}{\partial x}$ and $\frac{\partial z}{\partial y}$.

impact on the vibrating string problem.[44] Progress was not to come until another linear partial differential equation was studied in depth: Fourier's equation for heat diffusion,

$$K\frac{\partial^2 v}{\partial x^2}=\frac{\partial v}{\partial t}. \tag{1.26}$$

The equation is similar to the wave equation in that its form and linearity allow the possibility of solution by trigonometric series. But this time they were treated with great enthusiasm, and when Fourier put the initial conditions into them he obtained the kind of equation for which Lagrange had not been looking:

$$f(x)=\frac{2}{l}\sum_{r=1}^{\infty}\int_0^l f(X)\sin\frac{r\pi X}{l}\,dX\sin\frac{r\pi x}{l},\qquad 0\le x\le l, \tag{1.27}$$

$$f(x)=\frac{1}{l}\int_0^l f(X)\,dX+\frac{2}{l}\sum_{r=1}^{\infty}\int_0^l f(X)\cos\frac{r\pi X}{l}\,dX\cos\frac{r\pi x}{l},\qquad 0\le x\le l, \tag{1.28}$$

and

$$f(x)=\frac{1}{2l}\int_{-l}^{+l} f(X)\,dX + \frac{1}{l}\sum_{r=1}^{\infty}\left[\int_{-l}^{+l} f(X)\sin\frac{r\pi X}{l}\,dX\sin\frac{r\pi x}{l} + \int_{-l}^{+l} f(X)\cos\frac{r\pi X}{l}\,dX\cos\frac{r\pi x}{l}\right],\qquad -l\le x\le +l. \tag{1.29}$$

[44] The most important appearances seem to be, in chronological order:
1. Euler on infinite series in analysis (1729): the full series for terms with even multiples of the angle (Euler *1729a*, art. 55). The paper was not published until 1753.
2. Euler on the orbits of planets (1749): an interpolative argument towards the sine coefficients for the function $(1-g\cos\omega)^{-\mu}$, where the integral forms would have come by the procedure of taking the limit of a sum (Euler *1749a*, art. 29).
3. D'Alembert on the orbits of planets (1754): the constant term and the coefficient of $\cos x$ for the function $(a+b\cos z)^{n/2}$ (d'Alembert *1754a*, art. 232).
4. Clairaut on the apparent motion of the sun under the effects of perturbation (1754): the first n cosine coefficients for the function $(h-\cos t)^m$ (Clairaut *1754a*, esp. pp. 546–551).
5. Lagrange on finite trigonometric series, and their association with the vibrating string problem (1765): the coefficients for the first n sine terms, corresponding to his original n-body model (Lagrange *1765a*, art. 41). Riemann thought that when Lagrange wrote "$\int dX$" there (not "dx"!) he intended $\Sigma\Delta X$, that is, a summation of differences. But in this case Lagrange seems quite clearly to have intended the integral (see Riemann *1866a*, art. 2).
6. Euler on functions in analysis (1777): the cosine coefficients (Euler *1777a*: the paper was published posthumously in 1798).

In addition, Euler *1754a* contains series for various particular functions.

The details behind this achievement are too numerous to be given here;[45] but one point must be stressed. Fourier not only solved the problem of the calculation of these coefficients, *but also the deeper problem of periodicity.* It is from him that we have returned to geometry in the way that Euler hoped for. Lagrange was still alive in 1807 when Fourier presented his paper on heat diffusion, and was one of its examiners; and he raised such strong objections over these series that the paper was never published.[46] So deep in Lagrange's understanding of the vibrating string problem did the periodicity confusion lie that he failed to understand it even when Fourier explained it in the paper with clear diagrams, including three examples of functions set at zero over part of their interval, whose trigonometric respectability had explicitly been denied by Euler.[47]

"I am convinced" Fourier had concluded, "that the motion of the vibrating string is as exactly represented in all possible cases by trigonometric developments as by the integration which contains the arbitrary functions"—that is, the functional solution.[48] Lagrange objected strongly, possibly to Fourier's whole approach: not only to the method of separation of variables in (1.26) by means of the form

$$y = F(x)G(t), \tag{1.30}$$

which led to Bernoullian solutions in the first place (in flagrant contravention of his faith in functional solutions), but also to the series themselves. He also raised the technical question of convergence which had long been dormant. Fourier had contented himself with demonstrations for particular series, and now he replied to Lagrange with a special paper on the series

$$\frac{1}{2}x = \sin x - \frac{1}{2}\sin 2x + \frac{1}{3}\sin 3x - \cdots, \qquad -\frac{\pi}{2} \leq x \leq +\frac{\pi}{2}, \tag{1.31}$$

[45] See Grattan-Guinness *1969a*.

[46] Fourier *1807a*. The manuscript, along with selections from other unpublished manuscripts, is being prepared for publication by the author in collaboration with Dr. J. R. Ravetz as a survey of Fourier's life and work (see Grattan-Guinness and Ravetz *Fourier*). The circumstances of its rejection are discussed in Grattan-Guinness *1969a*, pp. 241–242, 250–251.

[47] Fourier *1807a*, arts. 63–74: the three examples are in arts. 70–72, and a quotation from Euler's denial in *1765b*, art. 22 (cited in n. 24, p. 9) is in art. 75.

[48] Fourier *1807a*, art. 77. It is worth remarking here that Fourier began his attack on heat diffusion in Lagrangian style by solving the problem for an n-body model and then taking n to infinity. Like Lagrange, Fourier missed the Fourier series because he was not looking for it, and found instead that when n was taken to infinity the t-term of his solution disappeared, thus implying a false steady-temperature situation in the corresponding continuous body. Only when he started completely anew by forming a partial differential equation for the continuous bodies did the infinite trigonometric series make their appearance (see Grattan-Guinness *1969a*, pp. 233–234, 243).

in which he found the form

$$-\frac{1}{2}\int_0^x \frac{\cos(n+\frac{1}{2})u}{\cos\frac{1}{2}u}\,du \tag{1.32}$$

for the remainder after n terms (n even), and showed that it tended to zero as n increased;[49] but in general he took the attitude of his predecessors in justifying the mathematics by means of its physical interpretation. All the solutions represented the temperature of a general point of some body at time t. The temperature was finite; therefore the series solution was convergent, which meant that in the particular case when $t = 0$ the Fourier series of (1.27)–(1.29) were convergent to their function $f(x)$.

Fourier's attitude was a consequence of his philosophical views on mathematics rather than a technical incompetence to handle convergence; as we shall see, he was to have considerable influence on the future development of the convergence problem of his series, one of the most important problems to be tackled in nineteenth-century analysis. For Lagrange, the last remaining pillar of the eighteenth-century analytical edifice, it was all simply unacceptable.[50]

[49] Fourier *1808a*.

[50] Four years after Fourier's paper Lagrange began to publish the second edition of his *Mechanique analitique*: his views on the vibrating string problem, of course, were basically unchanged. (See Lagrange *Mechanics*$_2$ (1811–15), part 2, sect. 6, esp. arts. 44–55. For the first edition, see n. 42, p. 17.)

THE BEGINNINGS OF NINETEENTH-CENTURY ANALYSIS

2

Lagrange died in 1813, and his faith in Taylor's series soon lost favor. The man who wrote most consistently on the new analysis was Augustin Louis Cauchy (1789–1857): thus a legend has grown that he swept all before him in a revolutionary program of new analysis which has remained standard ever since. Every aspect of this legend has veins of falsehood in it to match its truth. Some of the ideas that Cauchy developed expressed views which were held fairly generally in research circles at the time or had been introduced by others, although he never mentioned the fact; and many important features of the analysis which we ascribe to him did not come till a later generation.

Explanation of these points forms much of our later account: we set the scene now with the emergence of Cauchy as a mathematical force and the personal atmosphere within which his work was developed, beginning with a study in contrast between himself[1] and Fourier,[2] whose respective achievements were to come into remarkable conflict.

Cauchy was 20 years younger than Fourier. He was a zealous member of the Catholic and Bourbonist middle class, while Fourier was an orphan who built his career during the epoch of his leader and friend, Napoleon Bonaparte (1769–1821). Cauchy was a full-time professional mathematician who was sufficiently well-placed financially to resign his chairs and live independently—sometimes in self-exile from France—on the occasions when he refused to take oaths of allegiance demanded of him by various of the regimes which followed the end of the Bourbon period with the 1830 revolution. "One knows well enough what events made me wish to renounce the three chairs that I occupied in France," he wrote in the introduction to a paper written in Prague in 1835 after he had exiled himself from Paris in 1830, "and what majestic voice could alone determine me to relinquish the chair of mathematical physics that the King of Sardinia had deigned to entrust to me. But there is no doubt that close to the descendants of Louis XIV, close to those protector Princes, so en-

[1] The main biographical source on Cauchy is an excessively admiring life by his former pupil C. A. Valson (Valson *1868a*, vol. 1). J. Bertrand's review *1870a* of Valson's book and his own account of Cauchy's life *1904a* provide a very necessary antidote. B. Boncompagni's review *1869a* was a detailed synopsis of parts of the book, together with fragments of additional information found by Boncompagni himself. It is followed by a detailed list of all Cauchy's papers prepared by E. Narducci.

[2] The nearest approach to a biography of Fourier is Champollion-Figeac *1844a*. Jacques Champollion-Figeac was an archaeologist whose career was helped by Fourier at Grenoble, and so his story concentrates mainly on that period of Fourier's life and even devotes large sections to his own dealings with Napoleon rather than Fourier's. The book may be complemented by Cousin *1831a* and Arago *1838a*. The assessments in the text and footnotes following are based on these sources and the published works of the two men.

lightened in letters and sciences, I could believe myself capable of contributing to their progress."[3] The "majestic voice" was that of the dethroned Bourbon king, Charles X, who called Cauchy in 1833 from his chair of mathematical physics at Turin to educate his grandson. In 1838 he returned to Paris, but not to his professorships. His place at the Acadèmie des Sciences had been retained for him, however, and some payment went with it. The 1848 revolution brought Louis Napoleon to power and oath-taking was not required, so Cauchy resumed some of the teaching posts; but once again he resigned when Napoleon reinstituted oaths of allegiance with the founding of the Second Empire in 1852. Lengthy negotiations between both sides failed to achieve a compromise, until the authorities allowed him to teach again without taking the oath. In gratitude—or perhaps in return—Cauchy gave up the salary so earned to charitable purposes.

By contrast, Fourier was a part-time mathematician: the only period of his life that he spent in further education was from 1794 to 1798, when he was a student at the newly formed Ecole Normale in Paris, and then was appointed as *administrateur de police*, or assistant professor, at the Ecole Polytechnique to support the teaching of Monge on fortification and Lagrange and Laplace on analysis. After that he was burdened with heavy administrative duties: firstly as *secrétaire perpétuel* of the Institut d'Egypte and various other positions of responsibility during Bonaparte's Egyptian campaign (1798–1801), then as Prefect of the border department of Isère based at Grenoble (1802–1815), and finally at the Rhône Department around Lyons during Napoleon's second reign of "The Hundred Days" (1815). In fact, Napoleon's fall and the restoration of the monarchy left him temporarily without employment or income; but then he obtained the post of Director of the Bureau of Statistics of the Seine Department at Paris, which he kept for the rest of his life (1815–1830), and by 1822 his political fortunes had recovered so much from their post-Napoleonic depths that he was elected to the influential post of secrétaire perpétuel of the Académie des Sciences de l'Institut de France, which he also retained until his death.

"Cauchy is a fool, and one can't find any understanding with him, although he is the mathematician who at this time knows how mathematics should be treated . . . he is extremely catholic and bigoted. . . . I have

[3] Cauchy *1835a*, introduction; *Works*, ser. 2, vol. 10, pp. 189–190. The remarks were taken from *1833b*, a pamphlet which contained "a few words to men of good sense and good spirit." For much documentary information on Cauchy's period of exile in Turin in the early 1830s, see Terrazini *1957a*.

worked out a large paper on a certain class of transcendental functions to present to the Institut. I am doing it on Monday. I showed it to Cauchy: but he would hardly glance at it. And I can say without bragging that it is good. I am very curious to hear the judgement of the Institut. . . ."

So wrote Niels Hendrik Abel (1802–1829) in a letter of October, 1826, to his friend Bernt Holmboe (1795–1850) during his visit to Paris.[4] Well might he have been curious of the reception of his paper, which transformed the theory of elliptic integrals of Adrien Marie Legendre (1752–1833) into his own theory of elliptic functions! The tragic story of this manuscript is a good example of the Parisian scientific community at the time. Fourier, as secrétaire perpétuel, sent it to Legendre and Cauchy for examination. Cauchy took it first but he never looked at it: only when Carl Jacobi (1804–1851), who rivaled Abel in the theory of elliptic functions, noticed a reference to it in a later, published, paper by Abel and enquired of its whereabouts through Legendre after Abel's death did Cauchy return it to the Académie des Sciences with a grudging recommendation for publication. Even then, the 1830 revolution and the general inertia in the Académie delayed its appearance. Holmboe tried hard, but without success, to obtain it for his 1839 edition of Abel's works;[5] and only in 1841 was Guglielmo Libri (1803–1869) instructed to see it through the press. Libri possessed an enormous personal library, whose size was thought by some to have been created with the help of theft from public collections: he denied such allegations, but did not improve his reputation by "losing" Abel's manuscript just after the printing. This event helped to provoke a celebrated scandal, and the manuscript was rediscovered only in the 1950s in a Florence library.[6]

In contrast to Cauchy's selfish behavior, Fourier was always anxious

[4] Abel *Letters*, pp. 41 and 42; *Correspondence*, pp. 45 and 46. Cauchy was still alive when Holmboe prepared his edition of Abel's Collected Works in 1839, and so Holmboe tactfully omitted the remark on him in the selection of letters (see Abel *Works*$_1$, vol. 2, pp. 269–270). But it did appear in the second edition, edited by L. Sylow and S. Lie, in 1881 (see Abel *Works*$_2$, vol. 2, p. 259). Otherwise Sylow and Lie reproduced virtually unaltered Holmboe's excessively free translations of the letters to himself and to others which Abel had written in Norwegian. The most comprehensive source of Abelian correspondence is to be found in the *Memorial* published by Christiania (Oslo) University in 1902 to commemorate the centenary of his birth (Abel *1902a*). Both the Norwegian texts (Abel *Letters*) and (more accurate) French translations (Abel *Correspondence*) are included: all the translations used in this study have been made from the Norwegian originals by Madame Gerte Sinding-Larsen. (For a biography of Abel, see Ore *1957a*.)

[5] See Abel *Works*$_1$, vol. 2, p. viii.

[6] The paper appeared as Abel *1841a*, when Cauchy chose to make some appalling hypocritical remarks on it and its author in Cauchy *1841a*. On the history of the manuscript see Ore *1957a*, ch. 20; and on Abel's relations with the Académie des Sciences, Taton *1947a*.

for the welfare of others, from the resistance to the excesses of the Revolution which he showed during his early manhood, through the help he gave his people in Isère in his middle life, to the encouragement of young men such as Abel which he tried to offer, in spite of his frequent illnesses, during his last years. References in Cauchy's papers to the work of his contemporaries are frequently confined to people whose reputation was beyond reach, or to the chronology of rivals' papers, especially in the 1810s and 1820s when he was making his name as a great mathematician. Fourier did not give all the references he might have, either, but he was better liked by his colleagues. Cauchy was totally incapable of not publishing, or even of holding work back for a little while for possible modifications: at various periods of his life he even published journals comprised entirely of his own works,[7] and in his later years he inundated the Académie des Sciences with his papers to such an extent that a limitation of length of four pages had to be imposed on everybody else's contributions to its *Comptes Rendus*. In all, Cauchy wrote 789 papers, including 549 for the *Comptes Rendus* of the Académie des Sciences, and 141 in his own journals. Unfortunately, when the Académie des Sciences began to issue Cauchy's Collected Works in 1882, they did not make any use of the classification but followed their usual practice with papers of ordering them firstly by the journals in which they appeared and then chronologically for each journal; and Cauchy's own extremely vague cross-references were usually not supplemented by more detailed information. This method of editing, aimed at the minimization of editorial labor, is merely a nuisance in the handling of the works of, say Lagrange, Laplace, or Fourier; but in the case of Cauchy's excessively proliferous writing it is a positive handicap, since the proper understanding of his thought on a particular topic may require the location of perhaps a score of its appearances scattered throughout 26 large volumes. The value of the edition falls still lower when it is realized that notations have usually been "improved" to conform with contemporary usage, and that the volume devoted to papers published separately by Cauchy has still not appeared.[8] In contrast to this mass of publication, Fourier's heavy commitments outside scientific work meant that many of his projects remained

[7] Cauchy *1826–30a*, *1835a*, *1840–47a*.

[8] Valson complemented his biography (*1868a*, vol. 1) of Cauchy with a valuable classification of his output (vol 2: see also vol. 1, chs. 4, 9 and 10), and helped the Académie des Sciences with the edition; yet they did not use his classification, producing only an almost unusably complicated index of Cauchy's contributions to their journals in *Works*, ser. 1, vol. 12, pp. 466–504. The missing volume is ser. 2, vol. 15, and is due to appear soon; ser. 2, vol. 14 is erroneously titled as being a volume of papers published separately. See also Klein *1926a*, vol. 1, pp. 71–74.

unfinished and he left behind him an enormous mass of unpublished manuscripts.[9]

These were the two principal figures in early nineteenth-century pure mathematics, and no one was more aware of the fact than their common rival Siméon Denis Poisson (1781–1840).[10] Poisson, like Cauchy, was a professional mathematician—and a highly talented one—whose abilities would have received greater recognition had he not lived at the same time as greater men like Fourier and Cauchy. He competed with them on many issues of pure and applied mathematics; but the only advantage he could claim over them was more rapid publication of major papers in the Paris scientific journals, which has often caused commentators to overrate his achievements from the evidence of papers written under the influence of his rivals but published before their own work. By the 1820s his inferiority became so apparent that he lost even the advantages of quick printing, and his influence deteriorated considerably. Meanwhile, Fourier was fast recovering from his Napoleonic associations of 1815, and Cauchy was maintaining a high reputation by the quality of his work but creating personal enmities through his unpleasant personality.

So there was little friendship among them; but the situation must not be regarded as peculiar to these three, for rivalries, feuds and cliques were normal in French science at that time. Some (such as Cauchy and Poisson) participated more seriously than others (such as Fourier), but no one was unaffected. The reasons for such animosities were various, as our main characters indicate: personal antipathies and intellectual rivalries, strongly held political convictions in a violently changing political scene, and a tradition in French intellectual life for personal "competition." And these personal elements *do* play a role in the historical evaluation of their work, because they often influenced the style of its presentation and even its content. Paris was then the center of the scientific world; but the lines between the lines of intellectual brilliance were often packed with innuendo and double-meaning, with the most vicious attacks often coming when the victim was not mentioned by name. ". . . I don't think so much of the

[9] The principal source of Fourier manuscript is in the Bibliothèque Nationale in Paris: *Manuscrits Fonds Français*, volumes 22501–22529, totaling about 5200 sheets. They deal with problems in the theory of equations, inequalities, probability and errors of measurement, mechanics, statistics, friction, surface waves, heat diffusion, heat experiments, and terrestrial temperatures, and some are being used in the preparation of Grattan-Guinness and Ravetz *Fourier* (see n. 46, p. 20). Other less important sources of lecture notes and letters are to be found in various other public collections in France.
[10] For a biography of Poisson, see Arago *1850a*.

Frenchman as the German:" wrote Abel in his letter of October, 1826, "the Frenchman is very reserved with foreigners. It is very difficult to come into intimate contact with him. And I dare not count on that. Each one works for himself without caring what others do. Everybody wants to teach and nobody wants to learn. The most absolute egotism reigns everywhere. The only thing that the Frenchman seeks is to be practical: nobody can think outside himself. He is the only one who can produce anything theoretical. These are his thoughts, and that is why you can understand that it is difficult to be noticed, especially for a beginner."[11]

This is the atmosphere in which the new analysis was developed. We have already seen Fourier meeting opposition from Lagrange; now we find the young Cauchy entering into controversy with another member of the older generation, Legendre. At the start of his career Cauchy detected a problem in the writings of his predecessors which had had bearings on the foundations of eighteenth-century analysis: the use of complex numbers, especially in the evaluation of integrals. The problem occurred randomly and piecemeal; unlike the vibrating string problem, there was little concentrated discussion or development of ideas, but rather a common placid assumption that everything would be all right in the end. The attitude is to be found particularly in the writings of Euler, and also in the new *Exercises du calcul intégral*, then being published in installments by Legendre.[12] Legendre's work especially seems to have led Cauchy to the question: under what conditions is the use of complex numbers and variables valid in real-variable analysis?

This was the great problem of Cauchy's life: the ideas that he was to discover in his early years were to provide him and his contemporaries with problems of theory and application for decades, and to lead him to his great monument in mathematics—the theory of functions of a complex variable, and the calculus of residues to deal with their integration. It is not our intention here to give a detailed account of this exciting story, which itself would require a book for the telling,[13] but to indicate the effects on real-variable analysis that Cauchy's early work on it were to

[11] Abel *Letters*, p. 42; *Correspondence*, p. 46. Also partly in *Works*$_1$, vol. 2, p. 269; *Works*$_2$, vol. 2, pp. 259–260. We note the details of various rivalries, especially between Fourier, Cauchy and Poisson, at appropriate places in the text and in footnotes.
[12] Legendre *Exercises*.
[13] The book still needs to be written. For approximations to it, which tend to deal with particular aspects or periods only, see Casorati *1862a*, pp. 1–143; Brill and Nöther *1893a*, esp. pp. 155–202; Timchenko *1892–99a* (not seen); Stäckel *1900a*; Osgood *1901a*; Jourdain *1905a*; and Burkhardt *1908a*, pp. 671–745.

have. He was not the only mathematician of the time to be interested in the question: Poisson was trying to make progress with it, and the notebooks of Karl Friedrich Gauss (1777–1855) in Göttingen already contained ideas on its solution. But most of the ideas in Gauss's notebooks were to remain unpublished until long after his death. He could have been the hero of nineteenth-century analysis and indeed of pure mathematics in general. By 1814 his notebooks contained the seeds of most of the real-variable analysis and the problem of complex integration with which we are concerned, as well as possibilities for non-Euclidean geometry, linear algebra and elliptic functions. Yet he largely abandoned all these problems and left them to be developed anew by others, while he devoted the majority of his own energies to astronomy and geodesy, and later, to electricity and magnetism. Doubtless he felt a stronger attraction for problems of applied rather than pure mathematics (apart from his beloved number theory), and there is no crime in that; but it is almost a tragedy that he devoted especially his middle life to problems of geodetic surveys and the estimation of planetary orbits which relied far more on his colossal capacities for mental calculation than on his uniquely profound analytical gifts. He was a victim of his own abilities: had he not possessed computational as well as analytical genius he might have found these problems less attractive, if not insoluble, and so have returned to deeper work—to become King, rather than just Prince, of Mathematicians.[14]

Meanwhile it was Cauchy who achieved and made public the great progress on complex integration. His first paper was submitted to the Académie des Sciences in August, 1814 (during his 25th year), and was examined by Legendre and Sylvestre François Lacroix (1765–1843). Lacroix was the principal textbook writer of the period, and he possessed considerable historical and limited contemporary knowledge of mathematics. Hence his books fell below the standards of penetration set by Euler and Lagrange, tending instead to be compendia of established techniques moving into datedness; consequently, they were extremely popular in the general education of the time, appearing regularly in new editions throughout his life and even in posthumous editions revised by others.[15] Thus one cannot imagine that he held any strong views on

[14] The notebooks were first published in his Collected Works. We note relevant passages from these notes and also various works published by Gauss during his lifetime in later footnotes, and in the text. On Gauss's interest in complex integration, see Jourdain *1905a*, p. 205; and Brill and Nöther *1893a*, pp. 159–160, 169–170. For a general biography of Gauss, see Dunnington *1955a*.

[15] Lacroix's chief works on analysis were *Treatise* and *Calculus*. For information on his life and work, see Taton *1953a*.

Cauchy's manuscript on the evaluation of integrals; but with Legendre, interested in the same question, it was quite otherwise. Their report of the manuscript was submitted to the Académie des Sciences in November, 1814, and rightly showered praise on many of its features; but it also raised issue on certain results which were at variance with Legendre's own. Cauchy responded immediately with two supplements to the paper which seemed to satisfy the examiners, and they recommended publication in the *Mémoires présentés par divers savants.*

However, that was not the end of the story. Poisson, who had published in 1808 an insulting five-page summary of Fourier's 1807 paper on heat diffusion,[16] now performed the same service for Cauchy's 1814 paper at the end of that year,[17] but—typically—the journal to which the paper had been assigned failed to appear. Doubtless the political confusion following the fall of Napoleon in 1815 caused much delay; but when in the 1820s it was still not forthcoming Cauchy grew impatient, for both he and Poisson had been using and extending its results. In 1822 he hoped to publish it in the *Journal de l'Ecole Polytechnique*[18] and he did publish short extracts in other journals;[19] but, at last, the journal for which it had been originally destined was put in hand in 1825, and Cauchy sent the paper for publication together with the two supplements and the examiners' report, and new footnotes inserted in the text. Our consideration of the paper is based on this final form.[20]

Cauchy concentrated on "saving" complex variable integration in this paper by establishing conditions for validity of its operation. His procedure may be expressed as follows: let

$$z = h(x, y) + ik(x, y), \tag{2.1}$$

where h and k are single-valued real functions of the real variables x and y, and

$$f(z) = u(x, y) + iv(x, y), \tag{2.2}$$

where u and v are also single-valued real functions of x and y, and f is a single-valued function of z.

[16] Poisson *1808a*.
[17] Poisson *1814a*. Poisson himself was publishing papers on complex integration at this time (see Stäckel *1900a*, pp. 117–119).
[18] See the footnote in Cauchy *1823b*, p. 572; *Works*, ser. 2, vol. 1, p. 335.
[19] Cauchy *1822b* and *1826c*.
[20] Cauchy *1814a*. We use this date to emphasize the time of its composition. On the paper, see Yushkevich *1947a*, pp. 389–397.

Then Cauchy's condition for validity is

$$\frac{\partial^2}{\partial x\,\partial y}\int f(z)\,dz = \frac{\partial^2}{\partial y\,\partial x}\int f(z)\,dz, \tag{2.3}$$

since each side is

$$f(z)\frac{\partial^2 z}{\partial x\,\partial y} + f'(z)\frac{\partial z}{\partial x}\frac{\partial z}{\partial y}$$

and

$$f(z)\frac{\partial^2 z}{\partial y\,\partial x} + f'(z)\frac{\partial z}{\partial y}\frac{\partial z}{\partial x}$$

respectively. These last two expressions are seen to equal each other by an Eulerian appeal to the equality of the mixed differentials $\frac{\partial^2 z}{\partial x\,\partial y}$ and $\frac{\partial^2 z}{\partial y\,\partial x}$. Development of (2.3) gives

$$\frac{\partial}{\partial x}\left[f(z)\frac{\partial z}{\partial y}\right] = \frac{\partial}{\partial y}\left[f(z)\frac{\partial z}{\partial x}\right]. \tag{2.4}$$

From (2.1) and (2.2), (2.4) becomes

$$\frac{\partial}{\partial x}\left[(u+iv)\left(\frac{\partial h}{\partial y} + i\frac{\partial k}{\partial y}\right)\right] = \frac{\partial}{\partial y}\left[(u+iv)\left(\frac{\partial h}{\partial x} + i\frac{\partial k}{\partial x}\right)\right]. \tag{2.5}$$

Equating real and imaginary parts from each side of (2.5), we obtain

$$\frac{\partial}{\partial x}\left[u\frac{\partial h}{\partial y} - v\frac{\partial k}{\partial y}\right] = \frac{\partial}{\partial y}\left[u\frac{\partial h}{\partial x} - v\frac{\partial k}{\partial x}\right] \tag{2.6}$$

and

$$\frac{\partial}{\partial x}\left[u\frac{\partial k}{\partial y} + v\frac{\partial h}{\partial y}\right] = \frac{\partial}{\partial y}\left[u\frac{\partial k}{\partial x} + v\frac{\partial h}{\partial x}\right]. \tag{2.7}$$

A double integration with respect to x and y over (possibly variable) limits (x_1, x_2) and (y_1, y_2) yields finally

$$\int_{y_1}^{y_2}\left[u\frac{\partial h}{\partial y} - v\frac{\partial k}{\partial y}\right]_{x_1}^{x_2} dy = \int_{x_1}^{x_2}\left[u\frac{\partial h}{\partial x} - v\frac{\partial k}{\partial x}\right]_{y_1}^{y_2} dx \tag{2.8}$$

from (2.6) and

$$\int_{y_1}^{y_2}\left[u\frac{\partial k}{\partial y}+v\frac{\partial h}{\partial y}\right]_{x_1}^{x_2}dy=\int_{x_1}^{x_2}\left[u\frac{\partial k}{\partial x}+v\frac{\partial h}{\partial x}\right]_{y_1}^{y_2}dx \tag{2.9}$$

from (2.7).[21] These last two equations were to give Cauchy the mass of general theorems and special cases for the evaluation of integrals to which he was to devote the rest of his paper. Wisely he chose to corroborate some of the results of his predecessors as well as to produce many new ones. A particular example which he stressed immediately was the case

$$h(x, y)=x \quad \text{and} \quad k(x, y)=y. \tag{2.10}$$

This gives in (2.1) the standard complex variable

$$z=x+iy: \tag{2.11}$$

using it, the identities (2.6) and (2.7) become the "Cauchy-Riemann equations"

$$-\frac{\partial v}{\partial x}=\frac{\partial u}{\partial y} \tag{2.12}$$

and

$$\frac{\partial u}{\partial x}=\frac{\partial v}{\partial y}, \tag{2.13}$$

while (2.8) and (2.9) become

$$-\int_{y_1}^{y_2}\Big[v\Big]_{x_1}^{x_2}dy=\int_{x_1}^{x_2}\Big[u\Big]_{y_1}^{y_2}dx \tag{2.14}$$

and

$$\int_{y_1}^{y_2}\Big[u\Big]_{x_1}^{x_2}dy=\int_{x_1}^{x_2}\Big[v\Big]_{y_1}^{y_2}dx.^{22} \tag{2.15}$$

[21] Cauchy *1814a*, pp. 619–622; *Works*, ser. 1, vol. 1, pp. 336–339. Our use of z as a function of x and y follows current usage—Cauchy actually wrote y as a function of x and z.

[22] Cauchy *1814a*, pp. 622–623; *Works*, ser. 1, vol. 1, pp. 339–340. Equations (2.12) and (2.13) have been called the "Cauchy-Riemann equations" for the fundamental role that they play in Riemann's formulation of the theory of functions of a complex variable (see Riemann *1851a*, esp. art. 4). But it is clear that Cauchy saw the power of their application in 1814. (For eighteenth-century anticipations of the equations, see Stäckel *1900a*, pp. 113–115; and *1901a*, pp. 117–120.)

The problems of theoretical rather than evaluative interest were concerned with the validity of deriving (2.8) and (2.9) from (2.6) and (2.7) from the point of view of real-variable analysis itself. Both derivations are of the following form: if

$$\frac{\partial S}{\partial x}=\frac{\partial T}{\partial y}, \tag{2.16}$$

where S and T are functions of x and y, then

$$\int_{y_1}^{y_2}\int_{x_1}^{x_2}\frac{\partial S}{\partial x}\,dx\,dy=\int_{y_1}^{y_2}\int_{x_1}^{x_2}\frac{\partial T}{\partial y}\,dx\,dy; \tag{2.17}$$

and the difficulties arise in proceeding further to

$$\int_{y_1}^{y_2}\Big[S\Big]_{x_1}^{x_2}dy=\int_{x_1}^{x_2}\Big[T\Big]_{y_1}^{y_2}dx. \tag{2.18}$$

In the first place, a change of order is required on one side or the other of (2.17) to lead to (2.18) (on the right, in this case). Neither Cauchy nor the examiners, in the style of their time, suffered any qualms for the safety of such a procedure; but Cauchy did spot another problem, which was also usually taken for granted, for he raised the question of whether the double integrals obtainable from each of the orders of integration that were possible took on different values. He described the possibility, with justice, as a "singular property,"[23] and saw that it would arise from the occurrence of infinities in the integrands. Integrals of such functions he named as "singular integrals,"[24] which were to be evaluated by the now familiar formula

$$\int_a^b f(x)\,dx=\lim_{\varepsilon\to 0}\int_{a+\varepsilon}^b f(x)\,dx, \tag{2.19}$$

where the singularity occurs at $x=a$. Thus, if we write $F(x)$ for the indefinite integral of $f(x)$, (2.19) becomes

$$\int_a^b f(x)\,dx=F(b)-F(a+\delta), \tag{2.20}$$

where δ is "a very small quantity."[25] Cauchy then used (2.20) to compute

[23] Cauchy *1814a*, p. 665; *Works*, ser. 1, vol. 1, p. 381.
[24] Cauchy *1814a*, p. 678; *Works*, ser. 1, vol. 1, p. 394.
[25] Cauchy *1814a*, p. 674; *Works*, ser. 1, vol. 1, p. 391.

the difference in values between the two orders of integration of the double integral. If, for example, $\frac{\partial S}{\partial x}$ has a singularity at the point (X, Y) inside the rectangle bounded by $x = x_1$, $x = x_2$, $y = y_1$, and $y = y_2$, then he showed that

$$\int_{y_1}^{y_2} \int_{x_1}^{x_2} \frac{\partial S}{\partial x} dx\, dy - \int_{x_1}^{x_2} \int_{y_1}^{y_2} \frac{\partial S}{\partial x} dy\, dx$$

$$= \int_0^{\varepsilon} [S(X+p,\ Y+q) - S(X-p,\ Y+q) - S(X+p,\ Y-q)$$

$$+ S(X-p,\ Y-q)]\, dp, \tag{2.21}$$

and from (2.21) he developed his whole theory of complex variable functions and their integrals.[26]

We leave him now at the threshold of great things in the theory of complex variables, and return to the effects of his new ideas on real-variable analysis. These effects are summarized by (2.20); and they amount to nothing less than an attack on the hallowed inversion principle relating integration and differentiation. Put more generally, in the manner required for (2.21), a singularity in $f(x)$ at $x = c$ (where $a < c < b$) would cause the normal

$$\int_a^b f(x)\, dx = F(b) - F(a) \tag{2.22}$$

to be replaced by

$$\int_a^b f(x)\, dx = F(b) - F(a) - [F(c+\delta) - F(c-\delta)]. \tag{2.23}$$

Thus differentiation and integration no longer had the automatic relation to each other which had always been assumed; and the theory of functions was changing too, for not only did $f(x)$ assume infinite values but also its indefinite integral $F(x)$ showed corresponding discontinuities in the modern sense that Arbogast had emphasized. In fact, the problem was often more

[26] The main steps in Cauchy's argument are in *1814a*, pp. 672–678, 680–681, 691–704; *Works*, ser. 1, vol. 1, pp. 388–394, 396–397, 406–419. It is summarized in Jourdain *1905a*, pp. 190–200. One can judge Cauchy's impatience over the nonappearance of his paper by the superiority of presentation—especially from the point of view of involvement of functions of a complex variable—which he affected in the footnotes that he added in 1825 and also in the extended treatment of the whole subject in a paper published in that year (Cauchy *1825a*).

general still, for many of the "singular integrals" which Cauchy evaluated led to equations of the general form

$$\int_a^b G(k, x)\,dx = H(k). \tag{2.24}$$

Here k is a constant with respect to the integration, but otherwise it may be taken as another variable. Now the challenge to the inversion principle applies to G as a function of x; but the problem of functions appears not only for G in terms of x and k, but also for H with respect to k. Thus Cauchy became interested in functions not only in the conventional algebraic forms of his day, but also in *integral representations*; and it was with one particular integral representation that he met opposition from Legendre. It is defined by:

$$G(a, b, x) = \frac{x \cos ax}{\sin bx} \tag{2.25}$$

where a and b are constants, and

$$F(a) = \int_0^\infty \frac{x \cos ax}{\sin bx} \frac{dx}{1 + x^2} \tag{2.26}$$

where we regard b as fixed, and a as the new variable.

Cauchy had concluded from the application of his general relations (2.8) and (2.9) that

$$F(a) = \begin{cases} \dfrac{\pi}{2} \dfrac{e^a + e^{-a}}{e^b - e^{-b}} & \text{if } \ a < b, \qquad (2.27) \\[2ex] \dfrac{\pi}{2} \dfrac{2e^{-a} + e^{bs} - e^{-bs}}{e^b - e^{-b}} & \text{if } \ a > b, \qquad (2.28) \end{cases}$$

where

$$\frac{s}{2} = \frac{a}{2b} - n \tag{2.29}$$

and n is the integer nearest in value to $a/2b$.[27] In the examiners' report,

[27] Cauchy *1814a*, pp. 728–730, 759–762; *Works*, ser. 1, vol. 1, pp. 440–442, 469–472. Cauchy actually wrote (2.28) in two versions, according as $s > 0$ or $s < 0$.

Legendre expressed surprise that $F(a)$ was discontinuous of magnitude $\pi/2$ when $a = b$, $2b$, $3b$, . . . and especially mentioned to Cauchy "the inexactitude of his formula in the case of $a = b$."[28] He had in mind his own evaluation of $F(a)$ (by different means) when making this criticism; but his understanding of the situation was totally confused in a most interesting way. To show this, we begin with the even multiples of b:

$$a = 2nb, \quad \text{whence, from (2.29),} \quad s = 0. \tag{2.30}$$

Then (2.28) gives

$$F(2nb) = \frac{\pi}{2} \frac{2e^{-2nb}}{e^b - e^{-b}}. \tag{2.31}$$

But it is clear from (2.28) that $F(a)$ takes values close to $F(2nb)$ when a is close to $2nb$: in fact, $F(a)$ is differentiable there. Hence Legendre's description of $F(a)$ cannot be upheld in these cases.

However, for odd multiples of b (excluding, for the moment, $a = b$), equation (2.28) shows that there *is* a discontinuity in $F(a)$. The reason is that $a/2b$ is then of a form which may be expressed either as (some integer $+\frac{1}{2}$) or else as (next larger integer $-\frac{1}{2}$). We shall call this integer n_0 and consider the two cases as follows:

$$\text{Case 1.} \quad \frac{a}{2b} = n_0 + \frac{1}{2}; \text{ thus, from (2.29), } s = 1. \tag{2.32}$$

So from (2.28),

$$F(\overline{2n_0 + 1}\, b) = \frac{\pi}{2} \frac{2e^{-(2n_0+1)b} + e^b - e^{-b}}{e^b - e^{-b}}. \tag{2.33}$$

$$\text{Case 2.} \quad \frac{a}{2b} = (n_0 + 1) - \frac{1}{2}; \text{ thus, from (2.29), } s = -1. \tag{2.34}$$

In (2.28),

$$F(\overline{2n_0 + 1}\, b) = \frac{\pi}{2} \frac{2e^{-(2n_0+1)b} + e^{-b} - e^b}{e^b - e^{-b}}. \tag{2.35}$$

The two values of $F(\overline{2n_0 + 1}\, b)$ in (2.33) and (2.35) are clearly different; but the difference is easily seen to be π, and not $\pi/2$ as Legendre had suggested.

[28] Cauchy *1814a*, p. 609; *Works*, ser. 1, vol. 1, p. 326.

The same situation applies when $a = b$, although the proof is different. We have to use (2.27) as well as (2.28) (with $n = 1$ and $s = -1$) and we find from them the two equations:

$$F(b) = \frac{\pi}{2} \frac{e^b + e^{-b}}{e^b - e^{-b}} \tag{2.36}$$

and

$$F(b) = \frac{\pi}{2} \frac{3e^{-b} - e^b}{e^b - e^{-b}}, \tag{2.37}$$

again differing in value by π.

Why then did Legendre think that there was a discontinuity of only $\pi/2$ in $F(a)$ at every multiple of b instead of a discontinuity of π at the odd multiples? The answer to it is to be found in his own evaluation of $F(b)$ in the *Exercises du calcul intégral*, which he based on proving that

$$\int_0^\infty \frac{x}{m^2 + x^2} \frac{\sin 2bx}{1 + r^2 \pm 2r \cos 2bx} dx = \frac{\pi}{2} \frac{1}{e^{2bm} \pm r} \tag{2.38}$$

by expanding the second part of the integrand as a power series in r and then integrating the resultant series term-by-term. "These two formulae presume $r < 1$; but they are still true when $r = 1$," he had claimed optimistically, and obtained, for the lower pair of signs in (2.38),

$$\int_0^\infty \frac{x}{m^2 + x^2} \cot bx \, dx = \frac{\pi}{e^{2bm} - 1}.\text{[29]} \tag{2.39}$$

If we now put $m = 1$ in (2.39) we find that

$$\int_0^\infty \frac{x \cos bx}{\sin bx} \frac{dx}{1 + x^2} = F(b) = \frac{\pi}{2} \frac{2e^{-b}}{e^b - e^{-b}} \tag{2.40}$$

So we have *three* values for $F(b)$: two from Cauchy in (2.36) and (2.37), and one from Legendre in (2.40), the last being the arithmetic mean of Cauchy's two results. To Legendre, the "correct" value of $F(b)$ was, of course, his own; and so the behavior of young Cauchy's function $F(a)$ as it passed through $a = b$ was, in Legendre's words, to "increase or

[29] Legendre *Exercises*, vol. 2, p. 124. The date on the title page of the volume is 1817, but the installment containing these results had been published in June 1814, two months before Cauchy submitted his paper to the Académie des Sciences. (See Legendre's introduction to the volume.)

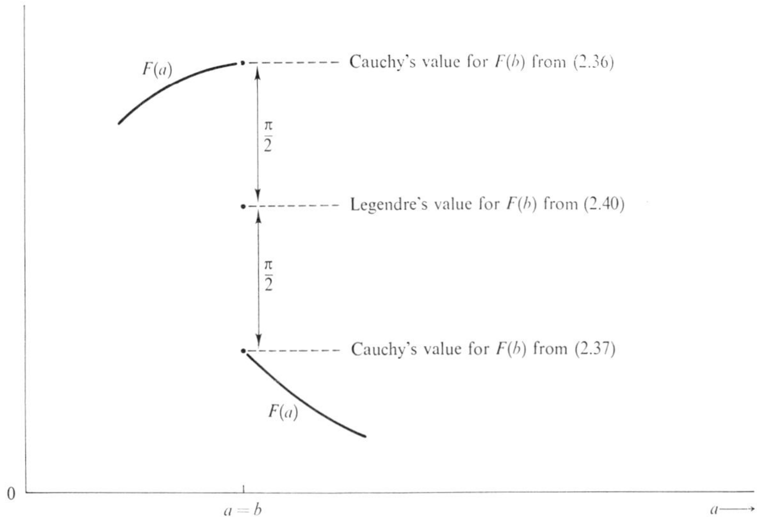

DIAGRAM 2.1

diminish suddenly by $\frac{1}{2}\pi$, when the ratio $\frac{a}{b}$ . . . diminishes or increases by an infinitely small quantity"[30] from the "correct" value which he himself had found (see diagram 2.1). He then must have presumed (incorrectly) that the same behavior is followed at all the other multiples of b.

The position is clear enough to our eyes: Cauchy's pairs of differing values for his function are actually left- and right-hand limiting values at the point in question. But our clarity is the consequence of just such difficulties as this one, for, as Cauchy wrote the two supplements to his paper, he had to face up seriously to the problem for the first time.

The first supplement was mainly concerned with explaining the need for the separate evaluation of $F(a)$ when $a < b$ and $a > b$ in the first place, while the second dealt with the evaluation of $F(b)$. Whether or not he had pursued an examination of his function similar to ours is not clear; at least he gave no signs of it in his reply, but instead produced a spurious reconciliation of his own value (2.36) with the "correct" value (2.40). The analysis is a classic of its genre and its time, and worth describing.

Put

$$a = b - \alpha, \tag{2.41}$$

[30] Cauchy *1814a*, p. 609; *Works*, ser. 1, vol. 1, p. 326.

where α is positive and infinitesimally small. This implies that its effects can be ignored in suitable places: for example, inserting (2.41) into (2.36), we learn that

$$F(b-\alpha)=\int_0^{\infty}\frac{x\cos(b-\alpha)x}{\sin bx}\frac{dx}{1+x^2}=\frac{\pi}{2}\frac{e^b+e^{-b}}{e^b-e^{-b}}, \tag{2.42}$$

and that it is (2.42) which needs conversion to the correct expression (2.40). This is done by means of the known results

$$\frac{\cos(b-\alpha)x}{\sin bx}=\cos\alpha x\frac{\cos bx}{\sin bx}+\sin\alpha x \tag{2.43}$$

and

$$\int_0^{\infty}\frac{x\sin\alpha x}{1+x^2}dx=\frac{\pi}{2}e^{-\alpha}, \tag{2.44}$$

which give

$$\int_0^{\infty}\frac{x\cos(b-\alpha)x}{\sin bx}\frac{dx}{1+x^2}=\frac{\pi}{2}e^{-\alpha}+\int_0^{\infty}\cos\alpha x\frac{x\cos bx}{\sin bx}\frac{dx}{1+x^2}$$

$$=\frac{\pi}{2}+\int_0^{\infty}\frac{x\cos bx}{\sin bx}\frac{dx}{1+x^2} \tag{2.45}$$

if once more discretion is shown in overlooking the presence of α,

$$=\frac{\pi}{2}+\frac{\pi}{2}\frac{2e^{-b}}{e^b-e^{-b}}$$

where the integral of (2.45) is chosen to be identified with the correct (2.40),

$$=\frac{\pi}{2}\frac{e^b+e^{-b}}{e^b-e^{-b}} \tag{2.46}$$

which is precisely the evaluation (2.42) of the integral, as required.[31] Similarly, one could show that (2.37) was also just the same as (2.40).

So Cauchy had done what was expected of him. He had practiced the craft of analysis in the style of his elders and had produced the kind of reasoning that they could understand: the proof of a desired result by the clever manipulation of algebraic expressions. But, also like his elders at

[31] Cauchy *1814a*, pp. 784–788; *Works*, ser. 1, vol. 1, pp. 493–496.

times, he had been *too* clever; for the whole argument, which completely satisfied Legendre,[32] is irrelevant to the problem at hand. For Cauchy, is was a baptism of fire into the foundations of contemporary analysis: yet he was not alone in his new territory, for Fourier had come across strikingly similar problems in his work on heat diffusion.

After rejecting Fourier's 1807 paper, the judges finally recommended a prize problem on heat diffusion for 1812. Fourier sent in a new paper in 1811 made up of a reworking of the old one (but omitting offensive diagrams and remarks!) as well as some new material, and it was crowned. But Lagrange was still unrepentant: the examiners' praise was qualified by his criticisms, and, as with Cauchy's 1814 paper, publication was considerably delayed. So Fourier began a third version in the form of a book, *La théorie analytique de la chaleur*, which was published, along with the 1811 paper, in the early 1820s.[33]

The two principal examiners of both Fourier's papers were Lagrange and Pièrre Simon Laplace (1749–1827).[34] Laplace was very much the politician in his scientific as well as in his public life, and when he saw the superiority of Fourier's achievements in heat diffusion over those of his own disciples Poisson and Jean Baptiste Biot (1774–1862),[35] he tended to shift his support in Fourier's direction.[36] Now, one problem left unsolved in the 1807 paper was that of heat diffusion in infinite continuous bodies. The lack of solution arose from a genuine difficulty in the periodicity of the trigonometric series, which allows the representation of a function to take place only over a finite interval. How then can a temperature be

[32] See his remark in Cauchy *1814a*, p. 609; *Works*, ser. 1, vol. 1, p. 326.

[33] The paper is Fourier *1811a*: the book is Fourier *1822a*. On these later developments see Grattan-Guinness *1969a*, pp. 232–233; Grattan-Guinness and Ravetz *Fourier*; and Jourdain *1917a*, p. 247. The examiners' report on the 1811 paper is quoted in Fourier *Works*, vol. 1, pp. vii–viii.

[34] Lacroix and Monge were the other examiners for the 1807 paper: in 1811 Monge was replaced by Haüy and Malus.

[35] We have already mentioned Poisson's insulting summary of Fourier's 1807 paper in n. 16, p. 31; by the time he published his own work on heat diffusion he had to admit defeat (see Poisson *1823a*, esp. pp. 1–2, 6). Biot more or less dropped the problem of heat diffusion: his early ideas had been of importance (for their influence on Fourier, see Grattan-Guinness *1969a*, pp. 234–235), but the pace that both Fourier and Poisson set was too fast for him, and his general treatise of 1816 on theoretical and experimental physics contained only a short section on it (see Biot *1816a*, vol. 4, esp. pp. 666–669).

[36] See the encouraging reference to Fourier's work in a paper of 1809 (Laplace *1809a*, p. 338; *Works*, vol. 12, p. 295), and in the last volume (1823–25) of his *Traité de mécanique celeste* (Laplace *Mechanics*, vol. 5, pp. 72–73; *Works*, vol. 5, second printing, p. 88).

expressed mathematically over an *infinite* range of values of the variable?

The problem is even more difficult than it sounds. We recall that Fourier was concerned with complete solutions to partial differential equations rather than merely with the special circumstances applicable when $t=0$; and not only did the periodicity problem arise then, but in some case the whole form of the solution itself collapsed.[37] So Fourier was at a standstill, unable to find the new solution form to replace his series.

Laplace took up the problem and in 1809 gave Fourier the clue for which he was looking. To Fourier's single-variable diffusion equation

$$\frac{\partial v}{\partial t}=K\frac{\partial^2 v}{\partial x^2} \tag{2.47}$$

with the initial condition

$$v=f(x) \quad \text{when} \quad t=0,\ -\infty \leq x \leq +\infty, \tag{2.48}$$

he found the solution

$$v=\frac{1}{\sqrt{\pi}}\int_0^\infty e^{-u^2}f(x+2u\sqrt{Kt})\,du.^{38} \tag{2.49}$$

So it was an *integral* form of solution which would solve the problem! Fourier was now on the scent, and to his 1811 paper he added a new section, containing integral solutions to the diffusion equation as given by the reciprocal relations

$$g(q)=\frac{1}{\sqrt{\pi}}\int_0^\infty f(p)\,{\sin \atop \cos}\,qp\,dp \tag{2.50}$$

and

$$f(x)=\frac{1}{\sqrt{\pi}}\int_0^\infty g(q)\,{\sin \atop \cos}\,qx\,dq, \tag{2.51}$$

which led to "Fourier's integral theorem":

[37] This happens with "nonharmonic" Fourier series; the point is explained in Grattan-Guinness *1969a*, pp. 249–250.

[38] Laplace *1809b*, pp. 235–244; *Works*, vol. 14, pp. 184–193.

$$f(x) = \frac{1}{\pi}\int_0^\infty f(p)dp \int_0^\infty \cos q(x-p)dq.^{39} \tag{2.52}$$

This was the work which brought Fourier and Cauchy closest together, for we see them both at work on integral representations of functions: indeed, the controversial integral (2.26) of Cauchy's 1814 paper *is* the Fourier cosine integral of the function $\frac{x}{(1+x^2)\sin bx}$. And the correspondence became even closer when Cauchy used integral methods of solving partial differential equations himself in a prize paper of 1815 on the motion of water-waves,[40] and then discovered Fourier's integral theorem for himself and published it in 1817.[41] The question of independent discovery is a difficult one: all of Fourier's work was still unpublished at the time, but the 1811 paper was with the Secretariat of the Académie des Sciences and therefore within Cauchy's reach. On the other hand, the rushed and excited character of the text of Cauchy's 1817 note suggests that it was his own work, and when Fourier acquainted him personally of his priority he published an acknowledgment in the following year.[42] So from then on, if not from an earlier time, it was clear that they were both working in the same problem areas; and a spirit of not entirely wholesome competition developed between the two of them (with Poisson dragging along behind), both with regard to methods of solving partial differential equations and also in foundational issues in real-variable analysis. The rising status of Fourier made Cauchy feel it politic to introduce acknowledgments to him in his writings during the 1820s. The 1815 water-waves paper, like the 1814 definite-integrals paper, did not appear until 1827 in the *Mémoires présentés par divers savants*; and the additions that he made to it in 1825 took the form this time of seven notes extra to the thirteen already placed at the end in the original version. The nineteenth note contained a more substantial acknowledgment to Fourier of the integral theorem, and another a reference to Fourier's introduction of the notation $\int_a^b$ to represent the definite integral. Fourier made the symbol popular in his book on heat diffusion: as a piece of notation it is an ingenious reinterpretation of Leibniz's summation symbol $\int$ for the indefinite integral as

[39] Fourier *1811a*, arts. 66–67, 71. In his book, see *1822a*, arts. 345–347 and 352–363, and the references to Laplace's result (2.49) in arts. 364 and 398.
[40] Cauchy *1815a*.
[41] Cauchy *1817a*.
[42] Cauchy *1818a*.

a "route-indicator" ${}_a\!\int^b$ of the passage of the variable x from a to b for the definite integral, and it immediately became standard.[43]

We have already seen Fourier's awareness of the problem of the convergence of his series, without an interest in seeking a general mathematical demonstration of the property. His attitude to the theory of functions and to integration, both for his series and his integrals, was similar: he saw the point clearly enough for his own purposes. He used the term "continuous" in its Eulerian sense and "discontinuous" to cover both the Eulerian and the modern senses, and claimed from specific examples that both types of discontinuous function were representable.[44] He also saw that such functions infringed the inversion principle, for he suggested an area interpretation of their integrals, especially for the integral forms (1.27)–(1.29) of the coefficients,[45] and proved his integral theorem (2.52) by taking the second integral as the limit of a sum.[46] On the other hand, he did not investigate at all a function which took an infinite value, because there would be no physical interpretation assignable to it.[47] It was Cauchy, with his closer contact with and deeper insight into these problems, and

[43] Fourier defined the notation in art. 231 of *1822a*, but in fact he used it first there in art. 222, and had first published it in *1816a*, p. 361. Cauchy's acknowledgment of it in the water-waves paper *1815a* occurs on p. 194, but *not* at the corresponding place in Cauchy's *Works*, ser. 1, vol. 1, p. 197. The widespread usage of the notation has been a trap for the editors of Collected Works. The editions for both Cauchy and Fourier updated all pre-$\int_a^b$ occurrences of definite integrals, which thus rendered rather peculiar all the explanations of the notation in the texts where it had actually been introduced for the first time; and for *1815a* the phrase:

> en désignant, avec M. Fourier, par la notation
>
> $\int_{\tilde{\omega}'}^{\tilde{\omega}''} f(\varpi)\, d\varpi$
>
> l'intégrale $\int f(\varpi)\, d\varpi$ prise entre les limites $\varpi = \varpi'$, $\varpi = \varpi''$

was omitted altogether between "par suite" and equation (22).

Other references by Cauchy to Fourier's invention may be found in an 1825 footnote to the 1814 integrals paper (Cauchy *1814a*, p. 623; *Works*, ser. 1, vol. 1, p. 340); a textbook of 1823 (Cauchy *1823a*, lecture 21; *Works*, ser. 2, vol. 4, p. 126); and a paper of 1823 on integrating partial differential equations (Cauchy *1823b*, p. 571; *Works*, ser. 2, vol. 1, p. 334).

The new notes added to Cauchy's water-waves paper include disguised accusations against Poisson of deception at the time of the original work (see Cauchy *1815a*, pp. 187–188; *Works*, ser. 1, vol. 1, pp. 189–191). On the rivalries between them and with Fourier in the field of partial differential equations, see Burkhardt *1908a*, pp. 423–426, 454–463, 476–477 (n. 2381), 495–501, 509 (n. 2528), and 511; Grattan-Guinness and Ravetz *Fourier*; and Grattan-Guinness *1970c*.

[44] For the series, see Fourier *1807a*, arts. 66–74, and the summarizing remarks in art. 75 (also in *1811a*, arts. 26–29 and 29; *1822a*, arts. 223–229 and 230). For the integrals, see Fourier *1811a*, art. 70; *1822a*, art. 358.

[45] Fourier *1807a*, art. 62 (with diagrams); *1811a*, art. 24; *1822a*, art. 220.

[46] Fourier *1811a*, art. 66; *1822a*, art. 346. See also arts. 415–418 for other uses of the geometrical interpretation of the integrals.

[47] See Fourier *1822a*, art. 417.

especially of the potentiality of the "singular integral" and the infinite value of the function in the integrand from which it arose, who took these developments much more seriously by writing his reconstruction of analysis and so creating the legend around his achievement which we must now begin both to record and to modify.

NEW IDEAS ON LIMITS

3

Around 1820 Cauchy began to rewrite in book form the courses of lectures in analysis which he had been giving at the Ecole Polytechnique and the other Paris colleges where he taught. The first of these works to appear was the most important one: the *Cours d'analyse de l'Ecole Royale Polytechnique* of 1821.[1]

The volume carried the subtitle *Première partie. Analyse algébrique*, and covered the principles of real and complex variable function theory. Although he did publish other textbooks (on differential and integral calculus), no further parts of the *Cours d'analyse* appeared, doubtless because of the pressure of research work and his self-imposed exile from France after the 1830 revolution. One may guess that they would have dealt with the solution of ordinary and partial differential equations, especially using the integral methods developed by himself and Fourier: if so, they would doubtless have been as replete with hidden meanings as was their predecessor. For textbook writing was not then the comparatively uncontroversial activity that it is today: the sharp distinction which now unfortunately exists between research work and teaching material was much less clear then, so that Cauchy's *Cours d'analyse*, like the textbooks of Euler and Lagrange before it, was well stocked with new results on problems which were still in a lively state of development. Given the personalities involved, it is therefore not surprising that rivalries manifest themselves in the *Cours d'analyse*, and in the silent way characteristic of the time.

The legend surrounding this book and its companions is that they revolutionized the whole of analysis and created the standards of mathematical rigor to which we are now accustomed. It is certainly true that they marked a great step forward from their predecessors; but there were also many crucial weaknesses whose solution by others was, in part, to create the standards of rigor of which we are speaking. That is a later part of the story; for the present we deal with the origins of the *Cours d'analyse* and the controversies it contained and created. These origins are more obscure than might be imagined, for parts of our thesis about Cauchy's analysis is that some of its ideas were borrowed without acknowledgment from the writings of others. No more direct statement can be made than this, precisely because Cauchy never named his sources: so our manner of presentation will be to state the results of the *Cours d'analyse* and the other books along with their alleged anticipations, and to give the reasons for thinking in such terms.

[1] Cauchy *1821a*.

The *Cours d'analyse* opens with an extraordinary introduction, whose main purpose is to announce the "program" of new analysis which is to follow. At the beginning it thanks Laplace and Poisson for urging Cauchy to prepare the book; at the end it thanks, among others, Poisson, for critical help; in between, it attacks enormous areas of Laplace's and Poisson's interests. There are three allusions to convergent and divergent series and the conditions of convergence, four to real and complex variables, and a paragraph on the pointlessness of applying mathematics outside its proper domain of natural science and on the uncertainty of the proof of Maclaurin's theorem which seems to be incomprehensible until we note Laplace's contemporary interest in probability and Poisson's frequent use of Taylor's series in analysis.[2] At one point Cauchy tells us that "in speaking of the continuity of functions, I could not dispense with making known the principal properties of infinitely small quantities, properties which serve as the basis of the infinitesimal calculus."[3] There is a double interest for us here which can serve as our introduction to the main body of the book: the use of infinitesimals and the formulation of the continuity of a function.

Both refer back to the 1814 paper on definite integrals, but the reference operates in different ways. We recall Cauchy's manipulations with infinitesimals there: the situation was to remain basically the same in the *Cours d'analyse* and indeed in the whole of his output, and was to cause his aspirations for analysis great setbacks. But on continuity there was an abrupt and utterly unexpected change of approach. The type of continuity normally applied at this time was Eulerian, identifiable with the good behavior of differentiable algebraic expressions. In speaking of "continuity" in his 1814 paper, Cauchy made no special reference or qualification to the term and presumably was using it in the orthodox way: indeed the theorems that he proved there would seem to require that restriction, in the way now familiar in the theory of functions of a complex variable. But in the *Cours d'analyse* we learn something very different:

[2] Cauchy *1821a*, introduction; see esp. pp. vi–vii. Laplace had just published new editions of both his treatise and his "popular summary" on probability (see Laplace *Probability*$_3$ (1820) and *Essay*$_4$ (1819)). Poisson's partiality for Taylor's series was probably a result of Lagrange's faith in it, although he did not become a disciple in the sense of Crelle and Arbogast (see n. 37, p. 15). One principal use was in the formation of partial differential equations in physical problems (see, for example, Poisson *1807a*, pp. 329–334).

[3] Cauchy *1821a*, introduction, p. ii.

DEFINITION 3.1
"The function $f(x)$ will remain continuous with respect to x between the given limits, if between these limits an infinitely small increase of the variable always produces an infinitely small increase of the function itself."[4]

This is the famous "Cauchy theory of continuity" which has remained standard ever since; we know that it embraces functions with corners as well as differentiable curves, such as are found with Fourier series. Now Cauchy's phrase "between the given limits" implies that he was concerned with specifying a range of values to a function, as had been emphasized by Fourier series. But it would be foolhardy to assume that Cauchy's new theory of continuity was introduced under the full influence of Fourier's results, or even of his own work on integrals. All the examples of continuous functions which he gave immediately following this formulation were of Eulerian algebraic expressions: indeed, since he had restricted it to finite-valued functions, he ruled that two of these examples—x^a with negative a, and a/x—were "discontinuous" at $x=0$, since there they became infinite.[5] It would appear rather that Cauchy was not extending the use of the term "continuity" to functions with corners, but *reformulating the old Eulerian algebraic sense in arithmetical terms.* The geometrical implications of this reformulation to cover functions with corners were *unintended* in 1821, and not realized until later. One cannot date the change of intent closer than to a paper of 1844, but then he published a retrospective account of the development of continuity which seems to have distorted the historical picture:

"In the works of Euler and Lagrange," wrote Cauchy, "a function is called *continuous* or *discontinuous*, according as the diverse values of that function, corresponding to diverse values of the variable ... are or are not produced by one and the same equation .... Nevertheless the definition that we have just recalled is far from offering mathematical precision; for the analytical laws to which functions can be subjected are generally expressed by algebraic or transcendental formulae [that is, by the Eulerian range of algebraic expressions], and it can happen that various formulae represent, for certain values of a variable x, the same function: then, for other values of x, different functions."

[4] Cauchy *1821a*, pp. 34–35; *Works*, ser. 2, vol. 3, p. 43.
[5] Cauchy *1821a*, pp. 36–37; *Works*, ser. 2, vol. 3, pp. 44–45.

As an example of a "misbehaving" algebraic expression he chose precisely the kind of integral representation with which he had struggled in 1814:

$$\frac{2}{\pi}\int_0^\infty \frac{x^2}{t^2+x^2}\,dt = \begin{cases} x & \text{if } x \geq 0 \\ -x & \text{if } x \leq 0. \end{cases} \tag{3.1}$$

In Euler's theory the left-hand side of (3.1) is "continuous" while the right-hand side is "discontinuous": "but the indeterminacy ceases if for Euler's definition we substitute that which I have given [in the *Cours d'analyse*]."[6]

But from where had the clever idea come in 1821? Cauchy does not say, and his own pre-*Cours d'analyse* gives not the slightest indication or motivation to it. But a possible source is to be found in the writings of the Bohemian scholar, Bernard Bolzano (1781–1848).

Like Fourier, Bolzano left behind him thousands of pages of manuscript; but in his case they covered problems not only in mathematics, but also in logic, religion, philosophy and ethics.[7] Evaluation of both his published and unpublished work in mathematics, if not all of his fields of interest, is far from complete; but our present purpose is merely to describe the mathematical results which had been published by Bolzano by the time of Cauchy's *Cours d'analyse*, and which could therefore have been available to him.

Firstly, there is the question of the circumstances of Bolzano's work. Bolzano lived his life in Prague, remote from the intellectual centers of the world: his mathematical interests were in the foundations of real variable analysis, Euclidean geometry, number theory, and rational and irrational numbers, and his work on these problems led him to ideas far ahead of his time. His writings were first brought to the general attention of the mathematical world by Herman Hankel (1839–1873) at the end of Hankel's life, when many of Bolzano's ideas were appearing in the current developments of analysis. Doubtless he also "spread the word" among his colleagues and contemporaries; but the discovery must have been fairly sudden for him, for in an address given two years previously, on mathe-

[6] Cauchy *1844b*, pp. 116–117; *Works*, ser. 1, vol. 8, pp. 145–146.

[7] These manuscripts reside mostly in libraries in Prague and Vienna. Parts of them are published from time to time (see Bolzano *Writings*, vols. 1 (analysis), 2 (number theory), and 5 (geometry); and *Numbers* (real numbers)), but in general they await their due editorial treatment. (For some account of Bolzano's life and work, see his autobiography *1836a* and also Winter *1949a*.)

matics during the last hundred years, Bolzano was not mentioned at all.[8] In any event, Bolzano's work rapidly became familiar to analysts at this time, and consequently it has been imagined that in Bolzano's own day nobody read him at all; but this is probably not true. His publications certainly did not achieve anything like the familiarity of the works of the Paris mathematicians, but they were known to some extent, especially in Germany, before and around the time of Cauchy's lectures. Abel, for example, was not only aware of them, but expressed his admiration in one of his Paris notebooks.[9] However, he did not make any reference to them in his own published works, and probably for a reason which applied to all of Bolzano's contemporaries and so contributed to the apparent ignorance of his output: his ideas were often so far ahead of their times that nobody could properly understand their purpose or develop them further. He issued his papers in the form of pamphlets, a common method of publishing scientific work at that time, and one of them was also inserted in the *Abhandlungen der königliche Böhmische Gesellschaft der Wissenschaften* in Prague. It would seem likely that Cauchy was aware of at least that paper and found it full of new ideas which he could use and extend in his analysis lectures.

The title of Bolzano's paper indicates its purpose: *Purely analytical proof of the theorem, that between any two values which guarantee an opposing result* [in sign] *lies at least one real root of the equation.*[10] The words that

[8] Hankel *1869a*; the references to Bolzano were in *1871a*, pp. 189, 209–210. Something of the significance of Bolzano's achievements in analysis may be gathered from Stolz *1881a*, pp. 255–268 (and corrections in Stolz *1883a*, pp. 518–519); Bolzano *1950a*, pp. 17–39 (by D. Steele); Jarník *1953a* and *1961a*; Kolman *1955a*, esp. chs. 2 and 3; Sebestik *1964a*; Wussing *1964a*; and Funk *1967a*.

[9] "Bolzano is a clever man," wrote Abel. (See Sylow *1902a*, pp. 6 and 13 (including a confession by Sylow that he had not heard of the man Bolzano before!); and Ore *1957a*, p. 96.) Abel hoped to meet Bolzano in Prague, but never did so (see Rychlik *1964a*).

Another mathematician to read Bolzano—in fact *1817a*—was Lobachevski (see Laptiev *1959a* and Folta *1961a*). Lobachevski's work outside non-Euclidean geometry is little known, but he read widely and wrote on many contemporary problems. His principal works on analysis were published in volume 5 of the excellent edition of his *Works*; but unfortunately they can take only this marginal place in our history, as they appeared (from 1834 onwards) in obscure Kazan journals, and, more importantly, with the exception of *1841a*, in Russian, a language known then by very few European mathematicians. His work does not make a substantial impact on foundational questions in analysis, and bears mostly upon the evaluation of Fourier integrals (including generalized kernels) by methods both of Fourier, and also of Cauchy and Legendre. (Commentary on this work is given in Lunts *1949a* and *1950a*; Gagaev *1952a*; and Paplauskas *1966a*, passim.)

[10] Bolzano *1817a*. The original German title is: *Rein analytischer Beweis des Lehrsatzes, dass zwischen je zwey Werthen, die ein entgegengesetztes Resultat gewähren, wenigstens eine reelle Wurzel der Gleichung liege.*

matter are "purely analytical," and indicate the forwardness of Bolzano's thinking in seeking a proof of that type rather than the geometrically founded arguments current at the time. Gauss had recently succeeded in using only analytical reasoning to show that every algebraic rational function could be decomposed into linear and quadratic factors:[11] now Bolzano was hoping to extend the powers of analysis to provide him with a proof of his own theorem. But he prefaced his proof with criticisms of the proof-methods of his contemporaries,[12] especially their practice of formulating continuity in a spatial and temporal framework, and replaced it by the following property of a finite valued function expressed in terms of the real number system:

DEFINITION 3.2

"... a function $f(x)$ varies according to the law of continuity for all values of x which lie inside or outside certain limits, is nothing other than this: if x is any such value, the difference $f(x + \omega) - f(x)$ can be made smaller than any given quantity, if one makes ω as small as one ever wants to; ..."[13]

Cauchy's formulation of continuity of 1821 is almost identical, even to the mention of "given limits" on the values of the variable;[14] but while Cauchy seemed to regard it only as a reformulation of old-style continuity, Bolzano had the extended interpretation in mind. He was acutely aware of the properties of functions and showed the difference between differentiability (Euler-continuity) and continuity in his own sense by later constructing a continuous nondifferentiable function, decades before the fashion for them developed at the time of Hankel's rediscovery of his work;[15] and in 1817 he was motivated toward consideration of continuity by the effort

[11] Gauss *1815a* and *1816a*.

[12] Bolzano *1817a*, preface, parts I–V.

[13] Bolzano *1817a*, preface, part IIa. Lacroix had given a vaguely similar formulation of continuity, but it seems unlikely to have been of use to Cauchy. (See Lacroix *Calculus*$_2$ (1806), art. 60.)

[14] They both also share a failure to specify positive values of the relevant expressions. This was done by Bolzano in a manuscript on analysis of the 1830s which has since been published. (See Bolzano *Writings*, vol. 1, p. 14.) For a summary of this manuscript, see Jarník *1931a*; and for later amendments to it made by Bolzano in a separate note, see van Rootselaar *1969a*.

[15] This is given in the 1830s manuscript (see Bolzano *Writings*, vol. 1, pp. 66–70, 88–89). The construction is described geometrically by the editor of the manuscript (K. Rychlik) in pp. 14–15 of his notes, and also in Jarník *1922a* and Kowalewski *1923a*. Hankel himself was interested in continuous nondifferentiable or infinitely oscillatory functions (see Hankel *1870a*). For specific examples of continuous nondifferentiable functions, see Weierstrass *1872a*; and Darboux *1875a*, sects. 7 and 8.

to prove the following generalization of the theorem of his title, which clearly applies to functions which are continuous in the wider sense:

THEOREM 3.1
If two continuous functions $f_1(x)$ and $f_2(x)$ satisfy the inequalities

$$f_1(\alpha) < f_2(\alpha) \quad \text{and} \quad f_1(\beta) > f_2(\beta), \qquad \alpha < \beta, \tag{3.2}$$

then there is at least one value a of x such that

$$\alpha < a < \beta \quad \text{and} \quad f_1(a) = f_2(a).\text{[16]} \tag{3.3}$$

The main body of the paper was devoted to the proof of this theorem, which incorporated a host of further new ideas on analysis which are also to be found throughout Cauchy's *Cours d'analyse*: the basic theorem itself (but not the generalization) found its way both into the main text with an intuitive geometrical proof, and also into one of the notes at the end of the book for those with a special interest in analysis.[17] There the proof given was a version of Bolzano's, and involved another important feature of his 1817 paper: limit-taking and the continuum.

The continuum is one aspect of analysis in which little common attitude can be found, for Bolzano eschewed Cauchy-like miracles with infinitesimals and, in another paper of 1817—on the rectification of curves and the calculation of areas and volumes—he explicitly excluded their use on the grounds of their lack of rigorous demonstration.[18] But on limits there was a remarkable accord, which is best explained against the background of the problem itself.

We take again Bolzano's definition 3.2 of continuity and write it as follows:

DEFINITION 3.3
$f(x)$ is *continuous* at $x = x_0$ if $[f(x +_0 \alpha) - f(x_0)]$ is small when α is small.

Let us reinterpret the definition as defining the limiting value $f(x_0)$ of $f(x_0 + \alpha)$ as α tends to zero, rather than continuity. It is continuity which guarantees that as a matter of fact this limit exists, and to avoid confusion

[16] Bolzano *1817a*, preface, part V; and art. 15.
[17] Cauchy *1821a*, pp. 43–44 and 460–462 respectively; *Works*, ser. 2, vol. 3, pp. 50–51 and 378–380.
[18] Bolzano *1817b*. The title includes the phrase "without consideration of the infinitely small . . . and any other assumption which is not rigorously proved." (In full, it is: *Die drei Probleme der Rectification, der Complanation und die Cubirung, ohne Betrachtung des unendlich Kleinen, ohne die Annahmen des Archimedes und ohne irgend eine nicht streng erweisliche Voraussetzung gelöst; zugleich als Probe einer gänzlichen Umstaltung der Raumwissenschaft, allen Mathematikern zur Prüfung vorgelegt.*) A note at the end of the preface of *1817a* indicated that *1817b* had just been completed.

we shall denote it by the symbol B (which is unconnected with $f(x)$) rather than $f(x_0)$. Then we have:

DEFINITION 3.4
The function $f(x_0+\alpha)$ has a (unique) *limit*, of *value* B, as $\alpha \to 0$ if $[f(x_0+\alpha)-B]$ is small when α is small.

In other words we may move as close as we wish to the limit B, *while still avoiding the limit itself.*

The full significance of Bolzano's definition 3.3 can only be grasped when seen in terms of the pattern of definition 3.4. Bolzano has defined continuity there, but he has done it in a limit-avoiding way in terms of arithmetical subtraction of expressions. This is the quintessence of his "pure analysis," for with it he started a revolution in approach to the subject: *the arithmetization of analysis by means of limit-avoidance techniques* such as in his definition 3.3. He used the approach consistently in his paper of 1817; and, remarkably, Cauchy suddenly started to use it when writing up his textbooks on analysis, including giving an explicit formulation of limit-avoidance in the preliminary chapter of the *Cours d'analyse*:

DEFINITION 3.5
"When the values successively attributed to a particular variable approach indefinitely a fixed value, so as to finish by differing from it by as little as one wishes, this latter is called the *limit* of all the others."[19]

We notice that the question of the use or non-use of infinitesimals has not arisen. The reason is that *the problem of the types of magnitudes admissible in analysis is logically independent of the problem of limits*: limit-avoidance or limit-achievement can be carried out in either an infinitesimal or a noninfinitesimal continuum. Yet in the development of the calculus from Newton and Leibniz onward it was usually thought that these two problems were linked, largely by the difficulties that had been found in formulating the derivative of the function

$$y=f(x) \tag{3.4}$$

by means of the equation

$$\frac{dy}{dx}=\lim_{\delta x\to 0}\left[\frac{f(x+\delta x)-f(x)}{\delta x}\right]. \tag{3.5}$$

[19] Cauchy *1821a*, p. 4; *Works*, ser. 2, vol. 3, p. 19. Lacroix had given a similar formulation of continuity (see his *Treatise*$_1$, vol. 1 (1797), p. 6), but did not develop its possibilities in the way that Bolzano did.

What happens when δx actually takes its limiting value 0? The ratio on the right-hand side of (3.5) becomes the indeterminate 0/0: so how is it that the derivative is calculated?

The problem appears even more sharply for many particular functions of x. For example, in calculating the derivative of

$$y = x^n \tag{3.6}$$

we make use of the binomial theorem

$$(x + \delta x)^n = x^n + nx^{n-1}\,\delta x + \frac{n(n-1)}{1 \cdot 2}\,x^{n-2}(\delta x)^2 + \cdots. \tag{3.7}$$

But this theorem only applies if $\delta x \neq 0$; hence, how is the undoubtedly correct answer

$$\frac{dy}{dx} = nx^{n-1} \tag{3.8}$$

obtained from (3.7)?

The attraction of infinitesimals to solve this problem was that, being smaller than ordinary numbers, they obeyed the law

$$a + h = a, \tag{3.9}$$

where a is the ordinary number and h the infinitesimal. Therefore they could be understood to be effectively zero and so allow the limiting value to be taken: on the other hand, being definitely different from zero they avoided the difficulty of the indeterminate 0/0. Hence they solved the problem of limits in a limit-*achieving* way, while at the same time avoiding that limit by being nonzero. The position is logically inconsistent, and in the early days of the calculus strenuous efforts were made to resolve the difficulties. But when the calculus began to develop in the eighteenth century under John Bernoulli and his pupils and then later with Cauchy and his contemporaries, the use of infinitesimals was mostly taken for granted: the vast and dignified corporation of results that had been built up using them was reason enough for confidence. Thus only occasional attention was given to foundational questions; for example, to Euler's laws (1.12) for the different orders of infinitesimals, to Lagrange's faith in Taylor's series and to d'Alembert's dislike of infinitesimals altogether. Bolzano not only distrusted infinitesimals but also *solved the problem of limits by introducing limit-avoidance*. Further, from what we have said, *limit-avoidance does not affect the use of infinitesimals*: nobody showed this better than Cauchy, who practiced limit-avoidance and infinitesimalism together. And when he came to the calculus he ran into a further difficulty that awaits the infinitesimalist there: the interpretation of $\frac{dy}{dx}$.

Let us return to our previous equation (3.8):

$$\frac{dy}{dx} = nx^{n-1}. \tag{3.8}$$

Bolzano's ideas show that we obtain this result by proving the limit-avoidance result:

$$\left| \frac{(x+\delta x)^n - x^n}{\delta x} - nx^{n-1} \right| \text{ is small if } |\delta x| \text{ is small,} \tag{3.10}$$

and he pointed out in his paper of 1817 that in a noninfinitesimal continuum, such as he was using there, $\frac{dy}{dx}$ acts only as a denoting symbol for the limiting value of the ratio (like the B in definition 3.4).[20] But if we use an infinitesimal continuum, $\frac{dy}{dx}$ seems to be more versatile. May we not multiply (3.8) through dx and deduce

$$dy = nx^{n-1}\, dx? \tag{3.11}$$

Infinitesimalists were in general not only happy with this procedure but regarded it as a bonus of the notation. But the procedure is invalid: what we *can* deduce from (3.8) is that the situation at an infinitesimally neighboring point, given by the increments dx and dy on x and y, may be described by the equation

$$dy = nx^{n-1}\, dx + q, \tag{3.12}$$

where q is a *second*-order infinitesimal quantity. Now q obeys the law

$$h + q = h, \tag{3.13}$$

where h is either an ordinary number or a first-order infinitesimal such as dx or dy; and it is the use of (3.13) in (3.12) which gives (3.11) from

[20] Bolzano *1817b*; see pp. xix–xxii of the preface, and arts. 1–9. The point is also made in his 1830s manuscript (*Writings*, vol. 1, pp. 80–89) and in his posthumous book on the *Paradoxes of the Infinite* (*1851a* (or *1950a*), esp. arts. 30–31 and 37).

(3.8).[21] The important point to realize is that (3.8) and (3.11) *are quite different in kind*: they are *not* equivalent ways of expressing the same situation. (3.8) informs us of the rate of change of the function (or slope of the tangent) at a particular point, while (3.11) is an equation relating two (infinitesimally) small increments of variable and function from that point. In (3.8), the notation $\frac{dy}{dx}$ is simply a denoting symbol for the derivative; in (3.11), it would be taken as an orthodox arithmetical quotient (and certainly not as 0/0!), as also in (3.12):

$$\frac{dy}{dx} = nx^{n-1} + \frac{q}{dx}. \tag{3.14}$$

So $\frac{dy}{dx}$ has a *double interpretation* in an infinitesimal continuum (following essentially from the double interpretation of the infinitesimal as a zero and a non-zero quantity), which does not apply in a noninfinitesimal field; for then all infinitesimal equations become logically meaningless, reducing to

$$0 = 0 \tag{3.15}$$

whether they are true or false in infinitesimals.

It is this double interpretation of $\frac{dy}{dx}$ which caused so much trouble to the users of the infinitesimal calculus. Cauchy is a good example. In his *Résumé des leçons données à l'Ecole Royale Polytechnique sur le calcul infinitésimal* of 1823, he expected the students to believe that if he put

$$i = \alpha h \tag{3.16}$$

where i and α, but not h, are infinitesimals, then since

$$\frac{f(x+\alpha h) - f(x)}{\alpha h} = \frac{f(x+i) - f(x)}{i} \tag{3.17}$$

[21] The philosopher Berkeley suggested that the derivative was calculated correctly because we make a pair of compensating errors in the course of the calculation (see Berkeley *1734a*, esp. arts. 21–25). But in working with infinitesimals he failed to notice that both errors are infinitesimally small of the second order and so behave like q in (3.12). More precisely, one of his errors is *exactly* q, and the other—concocted out of the difference between the alleged and actual lengths of the subtangent—while certainly numerically compensating the first, is irrelevant to the problem. For an analysis of Berkeley's ideas in terms of limit-avoidance and limit-achievement, see Grattan-Guinness *1970b*.

In the present study we do not discuss the important but quite separate question of the *consistent* construction of mathematical analysis in a system incorporating infinitesimals.

and therefore

$$\lim_{h\to 0}\left[\frac{f(x+\alpha h)-f(x)}{\alpha}\right]=h\lim_{i\to 0}\left[\frac{f(x+i)-f(x)}{i}\right], \tag{3.18}$$

he could deduce that

$$df(x)=hf'(x), \tag{3.19}$$

where $df(x)$ is the "differential" and $f'(x)$ the "derived function" (that is, the derivative) of $f(x)$. And he was not slow to exploit the mysteries of this remarkable equation: for by inserting into it the particular function

$$f(x)=x \tag{3.20}$$

he showed that

$$dx=h. \tag{3.21}$$

Thus the mysterious "differential" was made to take the finite value h on this occasion, and so when substituted into (3.19) (without justification, for (3.19) is a general equation but (3.21) applies only to the function x) gave the orthodox

$$df(x)=f'(x)\,dx \tag{3.22}$$

at last.[22]

More borrowing is certainly taking place in the course of this confusion —this time from Lagrange and Lazare Carnot (1753–1823). The notation $f'(x)$ and the term "derived function" are Lagrangian, the derivation taking place formally from Taylor's series;[23] but $df(x)$ and "differential" are due to Carnot, who defined the latter to be "the difference of two successive values of a particular variable when one considers the system to which it belongs in two or several consecutive states ... [it] indicates both that the quantity that it expresses is a difference and that this difference is an infinitely small quantity."[24] Carnot had an acute understanding of the workings of infinitesimals, which he developed to include an examination of different orders of infinitesimal quantities and the difference between d^2y and $(dy)^2$.[25] Cauchy's lack of acknowledgment to a

[22] Cauchy *1823a*, lecture 4.
[23] Lagrange *Functions*$_1$ (1797) and *Functions*$_2$ (1813), introductions.
[24] Carnot *Calculus*$_2$ (1813), arts. 46–47.
[25] Carnot *Calculus*$_2$ (1813), arts. 48–71. See also an attempt to use a theory of compensating errors à la Berkeley (see n. 21, p. 58) in arts. 3–11.

Napoleonic henchman like Carnot[26] is hardly surprising, but that he should have failed to understand the ideas of both Carnot and Lagrange to the extent of developing them both simultaneously is astonishing in a mathematician of Cauchy's abilities. His proof of the validity of differentiating a series term-by-term, which followed next, was almost comic. He felt the need of some more notation—this time Δx, which is the same as the i of (3.16) and therefore gives, from (3.16) and (3.21),

$$\Delta x = \alpha h = \alpha \, dx. \tag{3.23}$$

Now from the given equation

$$f(x) = u(x) + v(x) + w(x) + \cdots \tag{3.24}$$

it follows that

$$\Delta f = \Delta u + \Delta v + \Delta w + \cdots \tag{3.25}$$

(why?) and therefore

$$\frac{\Delta f}{\alpha} = \frac{\Delta u}{\alpha} + \frac{\Delta v}{\alpha} + \frac{\Delta w}{\alpha} + \cdots \tag{3.26}$$

or, from (3.23),

$$df = du + dv + dw + \cdots. \tag{3.27}$$

Dividing throughout by dx, we learn that

$$\frac{df}{dx} = \frac{du}{dx} + \frac{dv}{dx} + \frac{dw}{dx} + \cdots, \tag{3.28}$$

which, as Cauchy writes no more, is presumably the conclusion of the proof.[27] So perhaps Cauchy's calculus, with all its Carnot-style essays on

[26] After the fall of Napoleon in 1816, the new monarchy ordered the reconstruction of all the old Académies. Napoleon's close associates Lazare Carnot and Monge were omitted from the list of members of the Académie des Sciences, with Cauchy nominated to replace Monge. There was a measure of resentment against Cauchy amongst the members of the Académie des Sciences that he had not objected to the fact that his election was a consequence of a political act of vengeance against a dying man who had done more than anyone to re-create French higher education. (See Valson *1868a*, vol. 1, pp. 55–59; and Betrand *1904a*, pp. cxciv–cxcvi.)

[27] Cauchy *1823a*, lecture 5. In 1829, Cauchy published a revised and greatly extended version of the half of the *Résumé* which dealt with the differential calculus, under the title of *Leçons sur le calcul différentiel.* Unfortunately the intervening years and the greater space available did not seem to have improved his understanding of foundational questions (see Cauchy *1829a*, lectures 1 and 2).

For valueless remarks on Cauchy's contributions to the foundations of the calculus, see Dubbey *1966a*. More general studies of the analysis of this period on which we have not relied are to be found in Boyer *1949a*, chs. 6 and 7; and Manheim *1964a*, chs. 2–4.

different orders of infinitesimals,[28] is not as bristling with rigor as has been supposed.

The other aspect of the calculus is integration, where, by contrast, Cauchy made one of his most important contributions to analysis in interpreting the integral as an area rather than the inverse of differentiation. This was the logical development of his criticism of the inversion principle in the 1814 paper and in other work of that period.[29] In addition, Fourier's book on heat diffusion came out the year before Cauchy's *Résumé*, and Cauchy may have been anxious to improve on Fourier's bald statement that the integral was an area[30] to show that this area was the limiting value of the sum corresponding to a partition of the basic interval, and that it was independent of the manner of its achievement. His analysis showed his ability to handle both limits and continuity in the new way.

Cauchy confined his treatment of the integral to a finite continuous function $f(x)$ over a finite interval $[x_0, X]$. For the partition

$$x_0 < x_1 < x_2 < \cdots < x_{n-1} < X \tag{3.29}$$

of this interval he formed the sum

$$S = (x_1 - x_0)f(x_0) + \cdots + (X - x_{n-1})f(x_{n-1}) \tag{3.30}$$

(see diagram 3.1). His analysis relied on various theorems of the *Cours d'analyse*, of which the following (inspired by Bolzano?) was prominent:

THEOREM 3.2

Let $\{a_1, \ldots, a_n\}$ and $\{\alpha_1, \ldots, \alpha_n\}$ be two sets of real numbers, each set sharing the same sign among its members, and put

$$m = \min\{a_1, \ldots, a_n\} \tag{3.31}$$

and

$$M = \max\{a_1, \ldots, a_n\}. \tag{3.32}$$

Then

$$a_1\alpha_1 + \cdots + a_n\alpha_n = h(\alpha_1 + \cdots + \alpha_n), \tag{3.33}$$

[28] Cauchy *1821a*, ch. 2, art. 1; *1823a*, "Addition"; *1826d*; and *1829a*, lectures 3 and 6. In a later paper he tried to extend his ideas to functions of several variables and to partial differentiation (see Cauchy *1844a*, and the summary in *1843a*).

[29] See especially the successor to the 1814 paper which was written in the same year, but not published until 1844 (Cauchy *1844d*).

[30] Fourier *1822a*, art. 220.

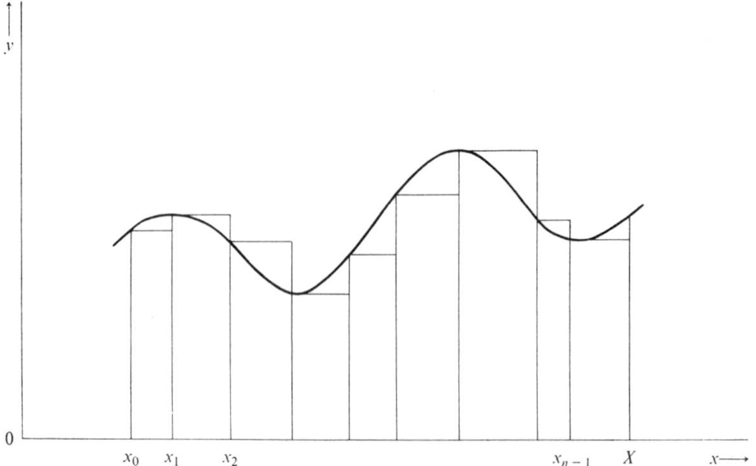

DIAGRAM 3.1

where

$$m < h < M.^{31} \tag{3.34}$$

If theorem 3.2 is applied to (3.30) we obtain

$$\begin{aligned} S &= [(x_1 - x_0) + \cdots + (X - x_{n-1})]h \\ &= (X - x_0)h, \end{aligned} \tag{3.35}$$

where h is within the maximum and minimum of $f(x)$ over $[x_0, X]$. By the continuity of $f(x)$ and the theorem which constituted the main aim of Bolzano's 1817 paper and was now in the *Cours d'analyse*, there would be at least one value of x within $[x_0, X]$ for which $f(x) = h$. Hence (3.35) could be written as

$$S = (X - x_0)f(x_0 + \theta(X - x_0)), \qquad 0 \leq \theta \leq 1. \tag{3.36}$$

A finer partition of $[x_0, X]$ could be taken to be a partition of each component of the original partition (3.29), and so, from (3.36), the corresponding sum S' may be written as

$$\begin{aligned} S' = (x_1 - x_0)f(x_0 + \theta_0(x_1 - x_0)) + \cdots \\ + (X - x_{n-1})f(x_{n-1} + \theta_{n-1}(X - x_{n-1})), \quad 0 \leq \theta_0, \ldots, \theta_{n-1} \leq 1. \end{aligned} \tag{3.37}$$

[31] Cauchy *1821a*, pp. 16–17; see also pp. 447–449. *Works*, ser. 2, vol. 3, pp. 28–29 and 368.

Since $f(x)$ is continuous, $f(x_{r-1}+\theta_{r-1}(x_r-x_{r-1}))$ will be close in value to $f(x_r)$ when the partition is sufficiently fine. Therefore we may write

$$f(x_{r-1}+\theta_{r-1}(x_r-x_{r-1}))=f(x_{r-1})\pm\varepsilon_{r-1}, \qquad r=1,\ldots,n, \tag{3.38}$$

and so (3.37) becomes

$$\begin{aligned} S' &= (x_1-x_0)[f(x_0)\pm\varepsilon_0]+\cdots+(X-x_{n-1})[f(x_{n-1})\pm\varepsilon_{n-1}] \\ &= S\pm(x_1-x_0)\varepsilon_0\pm\cdots\pm(X-x_{n-1})\varepsilon_n, \end{aligned} \tag{3.39}$$

from (3.35). By theorem 3.2, (3.39) becomes

$$S'=S+(X-x_0)h', \tag{3.40}$$

where h' is within the maximum and minimum of the ε_r. As the fineness of partitioning increases, the ε_r will tend to zero, and h' will decrease with them. So we can deduce from (3.40) that the effect on the sum S of subpartitioning the original partition will diminish as its fineness increases, and therefore that a limit will actually be reached.

To show that this limit was independent of the method of its achievement, Cauchy took (3.40) for two separate partitions to find that

$$S_1'=S_1+(X-x_0)h_1' \tag{3.41}$$

and

$$S_2'=S_2+(X-x_0)h_2'. \tag{3.42}$$

Then he formed the partition made up of the union of the original partitions. Since it could be taken as a subpartition of either of the originals, then its corresponding sum S_3 will apply in both (3.41) and (3.42), giving

$$S_3=S_1+(X-x_0)h_1' \tag{3.43}$$

and

$$S_3=S_2+(X-x_0)h_2', \tag{3.44}$$

and thus

$$S_1-S_2=(X-x_0)(h_1'-h_2'). \tag{3.45}$$

Therefore S_1 and S_2 can be made arbitrarily close to each other and so they both tend to the same limit.[32]

[32] Cauchy *1823a*, lecture 21.

Now began the reworking of the theory of integration: extension of the integral to functions with infinite values by means of the important idea of "singular integrals"

$$\int_a^b f(x)\,dx = \lim_{\varepsilon\to 0}\int_{a+\varepsilon}^b f(x)\,dx \tag{3.46}$$

of the 1814 paper,[33] to integrals over infinite intervals by means of a similar limit-taking technique:

$$\int_a^\infty f(x)\,dx = \lim_{b\to\infty}\int_a^b f(x)\,dx,^{34} \tag{3.47}$$

and verification of the basic properties of the integral. Among these, we may mention firstly the simple mean value theorem:

THEOREM 3.3
If $f(x)$ is finite and continuous over $[x_0, X]$, then

$$\int_{x_0}^X f(x)\,dx = (X - x_0)f(x_0 + \theta(X - x_0)), \qquad 0 \le \theta \le 1.^{35} \tag{3.48}$$

Next, we note the extended statement of the inversion principle:

THEOREM 3.4
If $f(x)$ is finite and continuous over $[x_0, X]$ then

$$\frac{d}{dx}\int_{x_0}^x f(t)\,dt = f(x) \tag{3.49}$$

and

$$\frac{d}{dx}\int_x^X f(t)\,dt = -f(x).^{36} \tag{3.50}$$

33 Cauchy *1823a*, lecture 25.
34 Cauchy *1823a*, lecture 24.
35 Cauchy *1823a*, lecture 22, equation (19).
In 1830 Galois made a contribution to analysis by proving a generalization of this theorem, in its differential rather than its integral form. For given functions $F(x)$ and $f(x)$, there exists a well-defined function $\phi(x)$ such that

$$\frac{F(x+h) - F(x)}{f(x+h) - f(x)} = \phi(k), \qquad x < k < x + h. \tag{1}$$

(See Galois *1830a*, art. 1.)
36 Cauchy *1823a*, lecture 26. Strictly speaking, theorem 3.4 is only half of the inversion principle: the other half, with $\int \frac{d}{dx}$ rather than $\frac{d}{dx}\int$ in (3.49) and (3.50), was taken for granted. Klein mistakenly said that Cauchy did not tackle the inversion principle in the *Résumé* (see Klein *1926a*, vol. 1, p. 85).

Finally, there is a theorem on integrating a convergent series term-by-term:

THEOREM 3.5

If $\Sigma_{r=0}^{\infty} u_r(x)$ converges to $s(x)$ for every x in the interval $[x_0, X]$, then we may integrate the series term-by-term; or, equivalently, we may interchange sum and integral. Symbolically, this may be written as follows. If

$$\sum_{r=0}^{\infty} u_r(x) = s(x) \tag{3.51}$$

over $[x_0, X]$, then

$$\sum_{r=0}^{\infty} \int_{x_0}^{X} u_r(x)\, dx = \int_{x_0}^{X} s(x)\, dx, \tag{3.52}$$

$$= \int_{x_0}^{X} \sum_{r=0}^{\infty} u_r(x)\, dx, \tag{3.53}$$

from (3.51).[37]

This last theorem involves another important aspect of Cauchy's reconstruction of analysis, where we find again limits and infinitesimals, and again success and failure: his theory of convergence of an infinite series. We also find again his *Cours d'analyse*—and Bolzano's 1817 paper on the roots of a continuous function.

[37] Cauchy *1823a*, lecture 40. The basic properties of the integral are discussed in lectures 23–33; see also a paper on differentiating under the integral sign (Cauchy *1827a*, esp. p. 125; *Works*, ser. 2, vol. 7, p. 160). The introduction of the *Leçons* of 1829 (see n. 27, p. 60) promised a similar revision of the part of the *Résumé* dealing with integration, but this was never written—presumably because of Cauchy's self-exile in 1830 (see Cauchy *1829a*, introduction).

SUCCESS AND FAILURE IN THE THEORY OF CONVERGENCE

4

"I am forced to refer the Taylor formula to the integral calculus," wrote Cauchy in the introduction to the *Résumé*. "I am not unaware that the illustrious author of the *Mécanique analytique* [that is, Lagrange] has taken the formula in question as the basis of his theory of derived functions. But, in spite of all the respect that such a great authority commands, the majority of geometers agree now to acknowledge the uncertainty of the results to which one can be led by the use of divergent series . . ."[1]

Lagrange's faith in Taylor series had several consequences, most of them unintended by him. Not only did it become clear that functions representable by a Taylor series did not embrace all the functions that were used in analysis, but also that the distinction between such functions and the rest involved the question of the convergence of the Taylor series itself, and hence of the convergence of infinite series in general. Lagrange himself assumed that the Taylor series was convergent, but in his later writings he groped uncertainly towards the forms

$$\frac{1}{(n-1)!}\int_0^h (h-t)^{n-1} f^{(n)}(x+t)\,dt \tag{4.1}$$

and

$$\frac{1}{n!}\,h^n f^{(n)}(x+\theta h), \qquad 0<\theta<1 \tag{4.2}$$

for the remainder of the series after n terms, which would have to decrease as n increases for convergence to take place.[2]

So far, so good; and it serves to indicate that eighteenth-century thinking on convergence was more than a blind belief that every series was convergent. The difficulties of the period were more subtle than that, and may be epitomized by a passage from a paper on divergent series written by Euler in the 1750s. Euler validly criticized the use of the geometrical progression to prove that

$$1+2+4+8+\cdots=\frac{1}{1-2}=-1, \tag{4.3}$$

[1] Cauchy *1823a*, introduction; *Works*, ser. 2, vol. 4, pp. 9–10. The passage is repeated in the introduction to the *Leçons* of 1829 (Cauchy *1829a*, introduction; *Works*, ser. 2, vol. 4, p. 267).

[2] Lagrange *Functions*$_2$ (1813), part 1, arts. 35 and 38–40 respectively. Still more primitive ideas are to be found in *Functions*$_1$ (1797), arts. 47–50. For commentary on Lagrange's remainders see Pringsheim *1900a*, pp. 440–450; and Brill and Nöther *1893a*, pp. 150–154. An improvement in derivation of the remainder form (4.2) was made in 1806 by Lagrange's pupil A. M. Ampère, who later achieved considerable distinction in electricity and magnetism (see Ampère *1806a*, esp. pp. 169–178; and also *1826a*).

replacing it by the accurate

$$1 + 2 + 4 + 8 + \cdots + 2^n + \frac{2^{n+1}}{1-2} = -1 \tag{4.4}$$

which implies the divergence of the series (4.3);[3] yet only a few pages later he was resorting to sophisticated finite difference methods to prove the seemingly similar result

$$1 - 1 + 1 - 1 + \cdots = \tfrac{1}{2}.^{4} \tag{4.5}$$

Why should there be this apparently complete change of intention in Euler's thought? The answer may be expressed as follows:

Eighteenth-century mathematicians interpreted an infinite series as a process of term-by-term addition of its members just as clearly as did their successors, and they were generally aware of the dangers of using divergent series in analysis. But they ran into unseen difficulties over the question of the method of summation. Term-by-term summation of an infinite series might be a technically awkward way of finding its sum, so more sophisticated methods were sought which would make summation easier. To take a well-known and long-established example, an easy way to find the sum

$$S = 1 + x + x^2 + x^3 + \cdots \tag{4.6}$$

of the geometric progression was to multiply the sum through by x to give

$$xS = x + x^2 + x^3 + x^4 + \cdots, \tag{4.7}$$

and then to subtract (4.7) from (4.6), yielding

$$(1 - x)S = 1, \tag{4.8}$$

and thus

$$S = \frac{1}{1-x}.$$

Euler's criticism of (4.4) was based on the insertion of the particular value 2 for x into (4.8) to deduce (4.3), on the grounds that the summation

[3] Euler *1754b*, arts. 4–8.

[4] Euler *1754b*, art. 15. Arts. 15 and 16 contain several summations of this kind, which are based on the "Euler transformation"

$$s = \tfrac{1}{2}u_1 - \tfrac{1}{4}\Delta u_1 + \tfrac{1}{8}\Delta^2 u_1 - \cdots, \tag{1}$$

where $\Delta^i u_1$ is the ith foward difference on the first term u_1, and the terms alternate in sign. Substantially the same analysis is to be found in the textbook on the differential calculus which Euler published at about the same time as this paper (see Euler *1755a*, part 2, ch. 1, esp. art. 9).

was not valid for that particular value of x. But he missed a deeper aspect of the problem: *the question of the nonorthodoxy of the method of summation itself.* We introduce a piece of modern terminology to explain the point:

DEFINITION 4.1
A method of summation of an infinite series, which is convergent according to the orthodox method of ordered term-by-term summation, is said to be *regular* if it gives the same sum for the series as does the orthodox method. Otherwise the method is said to be *irregular*.

Euler's difficulty with series was that *he assumed that all methods of summation were regular* and therefore gave "the" sum of the series: the fact that not only the sum of a series, but even the distinction between "convergence" and "nonconvergence," must be formulated *relative to the method of summation involved*, escaped the attention of both himself and his contemporaries. Therefore criticism was directed at objecting to the most glaring contraventions—such as (4.4)—of orthodox summation, without appreciating the mystery surrounding more subtle cases—such as (4.5). We may describe this situation as "sum-worship," in that the problem with an infinite series was felt to be merely to find its sum: if the sum was finite, then the series was "convergent," whereas if the sum was infinite it was "divergent." Euler was the high priest of sum-worship, for he was cleverer than anybody else at inventing unorthodox methods of summation (although he was often perfectly capable of considering orthodox convergence when treating certain particular series).[5] Therefore he suffered more than the rest from the bland assumption that they were equivalent to term-by-term summation; and the situation was aggravated by the fact that some of them were, so that the results obtained were sometimes right and sometimes wrong. Hence only fragments of the theory of (orthodox) convergence emerged during the eighteenth century. Euler, d'Alembert, Laplace and Lagrange all understood that the terms of a convergent series tended to zero as the series advanced,[6] and d'Alembert expressed the greatest doubt concerning any use of divergent series in analysis.[7] His warning appeared in a paper of 1768 on the binomial series

[5] A good example is the manuscript *Series*. Euler's correspondence, especially with C. Goldbach, often dealt with problems of methods of summation of series, with the convergence question largely taken for granted. See Fuss *1843a*; on convergence as a problem, see especially letter 83 of vol. 1 (Euler to Goldbach, August 7th, 1745). For a detailed study of Euler's achievements with infinite series, see Hofmann *1959a*.
[6] Euler *1754b*, art. 1; d'Alembert *1768c*, p. 177; Lagrange *1768a*, art. 37; Laplace *1779a*, pp. 207–208, and *Works*, vol. 10, pp. 1–2.
[7] D'Alembert *1768c*, p. 183.

$$(1+\mu)^m = \sum_{r=0}^{\infty} \frac{m(m-1)\cdot\cdots\cdot(m-r+1)}{1\cdot 2\cdot\cdots\cdot r}\mu^r, \tag{4.9}$$

where he even asserted that the series is convergent or divergent according as

$$\left|\frac{n\text{th term}}{(n+1)\text{th term}}\right| = \left|\frac{m-n+1}{n}\mu\right| \quad \text{is} < \text{ or } > 1. \tag{4.10}$$

But if this seems to have been a breakthrough by d'Alembert, then the remark of a few pages later that if $\mu = 99/100$ and $m = -2$ in (4.9) then the series is divergent up to the 99th term and convergent thereafter (because in (4.10)

$$\left|\frac{m-n+1}{n}\mu\right| = \frac{99}{100}\left(1+\frac{1}{n}\right)$$

and so is $>$ or <1 if $n <$ or > 99) shows that he was still distant from the real point.[8]

However, around the turn of the century changes began to take place. Lacroix's textbooks, Fourier's lecture notes and Gauss's notebooks all began to contain statements which focused attention only on the term-by-term summation of the series, and to formulate convergence and divergence (which would include what we now call "oscillatory" behavior) *solely* in terms of the behavior of the sum of the first n terms of the series, and prior to the question of the value of the sum.[9] Bolzano also made

[8] D'Alembert *1768c*, pp. 171, 176. This idea of d'Alembert is the origin of the term "d'Alembert's ratio test" which sometimes appears in later literature.

[9] Lacroix *Treatise*$_1$, vol. 1 (1797), pp. 4–9. Among the manuscripts of Fourier (see n. 9, p. 28) is a summary of lectures given presumably when he was at the Ecole Polytechnique between 1795 and 1798, in which he clearly stressed the need for term-by-term summation and the possibility of a series with no orthodox sum at all (see *Bibliothèque Nationale, Manuscrits Fonds Français* volume 22510, folio 65). Later he also gave evidence of his clear understanding of convergence in connection with Fourier series (see Fourier *1807a*, arts. 42–43; *1811a*, arts. 18 and 29; *1822a*, arts. 177 and 228). Gauss's notebooks (see n. 14, p. 30) include the manuscript *Series* from the period 1797–1814, which also stressed term-by-term summation and then classified series into categories beyond those of (orthodox) convergence and divergence. Another manuscript dealt with the summation of one of Euler's divergent series (Gauss *1797a*; compare Euler *1754b*, art. 13). Of his published work, see the allusion to convergence in his thesis (*1799a*, art. 6), and its formulation in a paper on the hypergeometric series (*1813a*, art. 15).

One may compare these three with R. Woodhouse, who wrote with admirable honesty on the problems which he failed to understand. One of these was convergence: his proposal for

$$1-1+1-1+\cdots \tag{1}$$

was to say that it summed to something called "$\frac{1}{1+1}$," which was to be distinguished from $\frac{1}{2}$. Of course, there is no harm in defining new symbols such as $\frac{1}{1+1}$; but the idea is "formalist" in the unflattering sense, and it does not bear on the problem of the convergence of series (see Woodhouse *1803a*, esp. pp. 58–63 and 160–162).

this point in his paper on the roots of a continuous function. Let us write $s_n(x)$ for the sum of the first n terms of $\Sigma_{r=0}^{\infty} u_r(x)$, and $r_n(x)$ for the remainder. Then

$$s_n(x) = u_0(x) + u_1(x) + \cdots + u_{n-1}(x) = \sum_{r=0}^{n-1} u_r(x) \tag{4.11}$$

and

$$r_n(x) = u_n(x) + u_{n+1}(x) + \cdots = \sum_{r=n}^{\infty} u_r(x), \tag{4.12}$$

and Bolzano stated as a "general theorem" that the convergence or nonconvergence of the series was to be judged only by the behavior of the sequence $s_n(x)$ as n increased.[10] Thus one may have series with an infinite sum, series which oscillate and especially "the class [of series] ... which possess the property that the variation (increase or decrease) which their value suffers through a prolongation [of terms] as far as desired remains always smaller than a certain value, which again can be taken as small as one wishes, if one has already prolonged the series sufficiently far."[11]

If we call this property of a series the "Bolzano property," then we may cast Bolzano's statement in the form of the following definition:

DEFINITION 4.2
The series $\Sigma_{r=0}^{\infty} u_r(x)$ possesses the *Bolzano property* if, given an order of smallness d, there is a value of n for which

$$|s_{n+r}(x) - s_n(x)| < d, \qquad r > 0.\text{[12]} \tag{4.13}$$

Now Bolzano stated the following theorem:

THEOREM 4.1
If the series has the Bolzano property, "then there always exists a certain constant value, and certainly only one, which the terms of this series always approach the more, and towards which they can come as close as desired, if one prolongs the series sufficiently far."[13]

In other words, the series is *convergent*, to a sum which we may write as $s(x)$. And Bolzano has phrased his theorem in a way *which again uses limit-avoidance*, this time in the context of the behavior of $s_n(x)$ for in-

[10] Bolzano *1817a*, art. 1. Bolzano wrote $\overset{n}{F}x$ for our $s_n(x)$.
[11] Bolzano *1817a*, art. 5.
[12] Bolzano gives this statement in *1817a*, art. 6, a passage which has been mistaken for his assertion of the necessity of his condition for convergence (see Bolzano *1950a*, p. 29; Sebestik *1964a*, p. 135; and Wussing *1964a*, p. 65).
[13] Bolzano *1817a*, art. 7.

creasing n, for from his property (4.13) he is trying to prove that: given an order of smallness d, there is a value of n for which

$$|s_n(x) - s(x)| < d. \tag{4.14}$$

But it is one thing to state a theorem and quite another to prove it. Bolzano's proof of theorem 4.1 was in two parts; the first part to show the existence of the sum-function, and the second, its uniqueness. Since the existence proof was prior to uniqueness, he assumed that he could take a variable value for the sum-function (varying, that is, independently of its being a function of x), which could therefore be selected to be within an arbitrary degree of closeness to what he called the "true value" of itself. If now an n were chosen for which

$$|s_{n+r}(x) - s_n(x)| < d, \qquad r > 0 \tag{4.15}$$

then, since

$$|s(x) - s_{n+r}(x)| < \omega \tag{4.16}$$

for some suitably small ω (why?), it follows from (4.15) and (4.16) that

$$|s(x) - s_n(x)| < d + \omega, \tag{4.17}$$

and so $s_n(x)$ is also arbitrarily close to the "true value" of $s(x)$ and therefore takes $s(x)$ as its limiting value. To show the uniqueness of $s(x)$, Bolzano took two different sums $s(x)$ and $t(x)$. Then

$$[s(x) - s_n(x)] < e_1 \tag{4.18}$$

and also

$$[t(x) - s_n(x)] < e_2, \tag{4.19}$$

whence

$$[s(x) - t(x)] < e_1 - e_2 \tag{4.20}$$

from (4.18) and (4.19), from which the equality of $s(x)$ and $t(x)$ follows.[14]

The proof of uniqueness is weak, although it can be repaired at once by the use of absolute values of the bracketed expressions. But the proof of the existence of $s(x)$ is quite faulty, for (4.16) simply begs the question: Bolzano had tried an arithmetical limit-avoiding proof for a theorem which

[14] Bolzano *1817a*, art. 7.

actually requires more than such techniques.[15] After such endeavors it is strange that he did not try to carry out the immeasurably easier proof of the *converse* of his theorem. For we have also the following:

THEOREM 4.2
If the series converges to the sum $s(x)$, then the Bolzano property (4.13) holds.

The proof is as follows: the assumption of convergence implies that for a suitably large n,

$$|s_n(x) - s(x)| < e_1 \tag{4.18}$$

and also

$$|s_{n+r}(x) - s(x)| < e_1. \tag{4.21}$$

Now

$$|s_{n+r}(x) - s_n(x)| < |s_n(x) - s(x)| + |s_{n+r}(x) - s(x)| < 2e_1 \tag{4.22}$$

from (4.18) and (4.21), which proves the Bolzano property as required.

Bolzano knew that his work was good; but he also knew that his living in the backwater of Prague prevented his work from receiving its due attention. So he ended the preface to his 1817 paper with a warning to others against the pursuit of reputation by the rapid publication of fragmentary ideas in large books, and gave the biography of the infrequent appearances of his own small mathematical tracts. And of his latest effort he pleaded to the scientific world: "... I must request ... that one does not overlook this particular paper because of its limited size, but rather

[15] Yet seeds of the still further new ideas are to be found in a later section of Bolzano's paper, although it would be unfair to have expected even him to have exploited them. From (4.13) we may deduce that

$$s_n(x) - d < s_{n+r}(x) < s_n(x) + d. \tag{1}$$

Now one of Bolzano's crucial steps en route to his main results was the following theorem: "If a property M does not apply to all values of a variable quantity x, but to all those which are smaller than a certain u: so there is always a quantity U which is the largest of those of which it can be asserted that all smaller x possess the property M." (Bolzano *1817a*, art. 12.) This theorem introduced into mathematics the idea of *the upper bound on a sequence*: a generalization of the result is that an infinite bounded set of points has an accumulation point. It was first proved by Weierstrass in his (unpublished) lectures at Berlin in the 1860s, and because of the theorem just stated, is known as the "Bolzano-Weierstrass theorem." (This name appears to have been given to it by Schwarz: see his *1872a*, p. 221 (footnote); *Papers*, vol. 2, p. 178. See also Cantor *1882a*, p. 114; *Papers*, p. 149.) The application of this theorem to the bounded infinite sequence of values $\{s_{n+r}(x)\}$ in (1) proves Bolzano's theorem 4.1, for the accumulation point corresponds to the sum function $s(x)$ whose existence he wished to prove.

examine it with all possible strictness and make known publicly the results of this examination, in order to explain more clearly what is perhaps unclear, to revoke what is quite incorrect, but to let succeed to general acceptance, the sooner the better, what is true and right."[16]

Cauchy, securely placed in the scientific capital of Paris, would seem to have answered Bolzano's plea in full measure, for his *Cours d'analyse* also contains Bolzano's theory of convergence in a fairly intact form.[17] One notices not so much the formulation of convergence (which was becoming generally understood at the time), but its presentation in terms of the Bolzano property, which itself was given now *as* the necessary and sufficient condition:

THEOREM 4.3
"For the series $[\sum_{r=0}^{\infty} u_r]$ to be convergent ... it is yet necessary that for increasing values of n the different sums

$u_n + u_{n+1}$,
$u_n + u_{n+1} + u_{n+2}$,
&c. ...

... finish by constantly achieving numerical values smaller than any assignable limit. Reciprocally, when these various conditions are fulfilled, the convergence of the series is assured."[18]

One notices further that Cauchy avoids the difficult proof by the bald statement that "the sums $s_n, s_{n+1}, \ldots$ differ from the limit s, and consequently among themselves, by infinitely small quantities."[19] And in the

[16] Bolzano *1817a*, preface, p. 28. The whole concluding section is on pp. 23–28.
[17] Cauchy *1821a*, pp. 123–131; *Works*, ser. 2, vol. 3, pp. 114–120.
[18] Cauchy *1821a*, pp. 125–126; *Works*, ser. 2, vol. 3, pp. 115–116.

In a paper on the harmonic series $\sum_{r=0}^{\infty} \frac{c}{a+rb}$, Euler showed that

$$\frac{(n-1)ic}{a+(ni-1)b} < s_{ni} - s_n < \frac{(n-1)ic}{a+ib} \tag{1}$$

and then advanced to a development of the "Euler constant"

$$\lim_{n\to\infty}\left[\sum_{r=1}^{n} \frac{1}{r} - \log n\right] \tag{2}$$

(Euler *1734a*, esp. art. 3). But it is optimistic of Eneström to see in this paper a Bolzano-Cauchy condition for convergence and divergence of the series based on the behavior of $|s_{ni} - s_n|$ as $i \to \infty$. As a condition, it is a special case of Bolzano's condition when r is a multiple of n; therefore it also is necessary. However, it is not sufficient, for as Pringsheim pointed out, it is refuted by $\sum_{r=2}^{\infty} \frac{1}{r \log r}$, which is divergent although $\lim_{i\to\infty} |s_{ni} - s_n| = 0$ then. (See Eneström *1905a* and Pringsheim *1905a*.)

[19] Cauchy *1821a*, p. 125; *Works*, ser. 2, vol. 3, p. 115. It was in the *Cours d'analyse* that Cauchy popularized the notations s_n, r_n, and s—and also "lim," which he may have learned from l'Huilier *1786a*, ch. 1.

notes at the end of the *Cours d'analyse* we find also a section on the real numbers system, including the remark from Bolzano's paper that an irrational number may be taken as the limit of a sequence of rationals. Bolzano had pointed out that in fact it cannot be assumed that the limiting value of a sequence of rational numbers is necessarily irrational, for the sum of the series $\Sigma_{r=1}^{\infty} (1/10)^r$, whose partial sums are all rational, is 1/9. Cauchy remarked that, in general, "when B is an irrational number, one can obtain it by rational numbers with values which are brought nearer and nearer to it."[20] The structure of the real numbers formed a significant part of Bolzano's later efforts to arithmetize analysis,[21] but it was of no importance in Cauchy's own analysis either before the *Cours d'analyse* or after it.

How may we sum up the situation between Cauchy and Bolzano? There is no documentary evidence of borrowing; but on the other hand, many circumstantial and internal factors strongly suggest a direct influence of thought. Of course, simultaneous independent discoveries are common enough in the history of science, but in this case we have to consider the origins of a *revolutionary new approach to analysis* based on the formulation of its fundamental components in terms of limit-avoidance, and the associated development of proof methods founded on the manipulation of arithmetical expressions. There is not the slightest hint of this revolution in Cauchy's work during the 1810s, which not only makes the coincidences of definition, condition and theorem seem especially frequent, but even the aspirations of the *Cours d'analyse* themselves surprising. The question of how Cauchy came across Bolzano's paper is impossible to answer, but the possibility of him doing so is not difficult to envisage. Paris, the scientific Mecca of the age, must have received nearly all the current literature by some means or other, and the professor of mathematics at the Ecole Polytechnique, the Sorbonne, and the Collège de France, who was also a member of the Académie des Sciences, would have had ample access to it. For example, the Bibliothèque Impériale (now the Bibliothèque Nationale) began to take the Prague journal *with*

[20] Bolzano *1817a*, art. 8. Cauchy *1821a*, pp. 409 and 415; *Works*, ser. 2, vol. 3, pp. 337 and 341.

[21] Bolzano's theory of rationals and irrationals remained in manuscript until recently (see Bolzano *Numbers*), but the problem had been yet another field of interest for Hankel's contemporaries in the 1870s. The accuracy of his theory is disputed (see Rychlik *1961a*, van Rootselaar *1964a* and Laugwitz *1965a*). Bolzano's manuscript included a reworking of some of his ideas on analysis, including his sufficient condition for convergence of a series and his version of the Bolzano-Weierstrass theorem (Bolzano *Numbers*, arts. 102 and 104).

precisely the volume which contained Bolzano's 1817 paper: it is not beyond belief that Cauchy turned to the new periodical to see if there was anything of interest in it for him.[22] His writings quite often reflect a detailed knowledge of current literature, especially of those parts of it which contained views consonant with, or attributive to, his own. When he came to deliver his analysis lectures during the late 1810s, he was looking around for ideas. Those which he undoubtedly took from Lagrange and Carnot, were taken without understanding; but in Bolzano's paper he found the perfect outlet for his particular genius. For it is a characteristic of Cauchy's mathematics that the basic stimuli come from elsewhere, and then are developed far beyond their original bounds. His analysis, if seen as inspired by Bolzano's fragments, follows this pattern perfectly; for, having learned from Bolzano how to reinterpret and reformulate the basic components of analysis in terms of limit-avoidance, he built up an enormous superstructure of theorems and properties. That part of the Cauchy legend is true: we shall see that some of its crucial members were faulty, but much of it was sound, and one part—dealing with the devising of tests for the convergence or divergence of a series—was so successful that we have had to devote an appendix solely to its description and to the aftermath that it inspired. No priority row sprang up over Cauchy's *Cours d'analyse*, although Bolzano was later aware of the book;[23] but then that quiet Bohemian scholar was not the man for such things, seeing

[22] The journal was also available in Paris at the library of the Muséum National d'Histoire Naturelle, but this would have been an unlikely source for Cauchy. It does not seem to have been possessed either by the Académie des Sciences or by any of the institutions at which he taught. As a footnote to the relationship between Cauchy and Bolzano, they both lived in Prague during the period 1833–1835. According to Struik and Struik *1928a*, they did not meet; but Bolzano was looking forward to meeting Cauchy (see the letter to Přihonský of August, 1833 in Winter *1956a*, p. 156) and a recently published letter by Bolzano in Seidlerová *1961a* shows that there *were* meetings, in the spring and autumn of 1834, when Bolzano gave Cauchy a manuscript on the rectification of curves and later suspected—in the end groundlessly—that Cauchy had used these ideas in some of his own work on the subject! (For quotations and further details of this affair, see Grattan-Guinness 1970*c*.) It is clear from Cauchy's contributions to the meetings of the Académie des Sciences and from references in his writings that he knew German.

[23] In Bolzano's 1830 manuscript, he gave a list of occurrences of the arithmetical formulation of continuity in current literature, including those in Cauchy's *Cours d'analyse* and the pre-Bolzano form of Lacroix mentioned in n. 13, p. 53 (Bolzano *Writings*, vol. 1, p. 15; see also p. 94). Another of his references was to Martin Ohm, the younger brother of the physicist Simon Ohm. Ohm was the Lacroix-style textbook writer in Germany, and, like Lacroix, mainly reported current developments (in his case, Cauchy and after) rather than made contributions to them. (See especially Ohm *Mathematics*; Bolzano's reference is to vol. 2 (1829), art. 456. Probably Ohm was following Cauchy.)

in the book only the answer to his plea for attention, or possibly not even thinking that the common ideas were entirely of his own invention. For, needless to say, the name of Bolzano appears nowhere in the *Cours d'analyse*: Cauchy would have had more sense than to make Bolzano's work known to his rivals. If one doubts his ability to omit references to those whose work he is discussing or challenging, then the first theorem that he was to state in the chapter on convergence should dispel them:

THEOREM 4.4
"When the different terms of the series $[\Sigma_{r=0}^{\infty} u_r]$ are functions of the same variable x, continuous with respect to that variable in the vicinity of a particular value for which the series is convergent, the sum s of the series is also a continuous function of x in the vicinity of that particular value."[24]

First, a word about the term "continuous." It is, of course, used in the sense of the new formulation, but whether or not Cauchy had in mind its extended interpretation is uncertain. In any case, either interpretation would lead to the crisis he intended—Fourier series. For the sine and cosine functions are continuous, even in the narrow sense, and Fourier series are infinite series of such functions; therefore, according to the theorem, the Fourier series of any discontinuous function is not convergent to it.

This is the purpose of Cauchy's theorem, placed as it is in the *Cours d'analyse* as the first theorem on infinite series which the student should learn. Cauchy was challenging the whole principle of Fourier series; but in mentioning not even Fourier's name[25] or work, quite apart from the important clash of ideas presented by Fourier's series for discontinuous functions and his own theorem, he must be regarded as guilty of less than professional conduct as a teacher. For Cauchy did support his own position with a proof, which we now describe.

Throughout his treatment of convergence, Cauchy made no distinction between series of constant terms and series of functions, and wrote

$$s = u_0 + u_1 + u_2 + \cdots \tag{4.23}$$

$$= s_n + r_n, \tag{4.24}$$

where s_n is the nth partial sum and r_n the remainder. His proof of theorem 4.4 may be divided into the following steps:

[24] Cauchy *1821a*, pp. 131–132; *Works*, ser. 2, vol. 3, p. 120.
[25] Fourier's name does appear once in Cauchy's *Cours d'analyse*; but only in one of the notes at the end, and in connection with their common interest in the roots of equations. (Cauchy *1821a*, pp. 504–505; *Works*, ser. 2, vol. 3, p. 412. The paper mentioned by Cauchy is Fourier *1818a*.)

1. s_n is continuous for a finite n.
2. Therefore, for an (infinitely) small increment α of x, the increment on s_n is also infinitely small.
3. Since, by assumption, the series is convergent at $x = x_0$, r_n is small and the increment on it due to α "becomes insensible at the same time" as r_n.
4. From (4.24), the increment on s equals the sum of the increments on s_n and r_n.
5. From steps 2 and 3, these two latter increments are small. Therefore, from step 4, the increment on s is also small, and so s is continuous, as required.[26]

We shall postpone the resolution of the controversy: now we record the first public mention of it. This occurred in a footnote to a paper of 1826 by Abel, which contained his own great contribution to the development of analysis.

Abel was the first great convert of Cauchy's *Cours d'analyse*; but before he saw the light he practiced pre-Cauchian analysis in many of its features, and liberally stocked his juvenilia with Eulerian summations of series.[27] He came to Berlin in the winter of 1825–26 and met August Leopold Crelle (1780–1856), a self-taught mathematician and brilliant engineer who made a fortune from construction projects of various kinds and was planning to spend some of it on the founding of a new journal in mathematics. Crelle's own writings on mathematics seem to have been of but slight importance, but he more than compensated for this by the insight into the abilities of others which successful businessmen often develop: when he saw Abel's work he realized that he had available an author of sufficient caliber to give the prospective journal the impact that it would need, and so he began to publish the *Journal für die reine und angewandte Mathematik* in 1826. In fact, soon after commencement of publication, he offered to stand down as editor in favor of Abel, but Abel declined the offer.[28]

Crelle benefited from Abel's genius: and on the other hand, Abel benefited from Crelle's considerable library, within which he was given complete freedom and where he found the latest works, especially Cauchy's *Cours d'analyse*. Soon his youthful malpractices with series were behind

[26] Cauchy *1821a*, p. 131; *Works*, ser. 2, vol. 3, p. 120.
[27] See Sylow *1902a*, pp. 11–12.
[28] See the letter from Abel to Hansteen of March 29th, 1826, in Abel *Letters*, p. 22; *Correspondence*, pp. 23–24. For Crelle's reminiscences, see his *1829b*.

him: he became more Cauchian than Cauchy. "Divergent series are on the whole devil's work," he wrote to Holmboe in January 1826, "and it is a shame that one dares to found any proof on them. One can get out of them what one wants if one uses them, and it is they which have made so much unhappiness and so many paradoxes. Can one think of anything more appalling than to say that

$$0 = 1^n - 2^n + 3^n - 4^n + \quad \text{etc.}$$

where n is a positive number. Here's something to laugh at, friends. My eyes were opened in a very astonishing manner ... there does not exist in the whole of mathematics one infinite series whose sum is defined in a strict way: in other words, the most important thing in mathematics stands without foundation . . . even the binomial formula is not well proved. . . ."[29]

Abel's remarks epitomize the new attitude to convergence: an excessive reaction against Euler's faith that all methods of summation were regular into a faith that they could not be. The attitude was to dominate mathematics for the rest of the century: not until the 1890s were nonorthodox methods and their associated theories of convergence and nonconvergence studied alongside, but apart from, the orthodox theory established by Bolzano and worshipped by Abel.[30]

The reference to the binomial series in his letter to Holmboe is symptomatic of his approach to analysis. He was always fascinated by power series,[31] and around the time of that letter he followed d'Alembert in making a special study of the binomial series

$$1 + mx + \frac{m(m-1)}{1.2} x^2 + \cdots$$

[29] Abel *Letters*, p. 16; *Correspondence*, pp. 16–17. Mostly also in Abel *Works*$_1$, vol. 2, pp. 266–267; *Works*$_2$, vol. 2, pp. 256–257.

[30] The systematic investigation of nonorthodox methods of summation was begun by Borel in 1895—and he became interested in the problem after reading the introduction to Cauchy's *Cours d'analyse* on convergence and divergence! (See Collingwood *1959a*, p. 493.) But the Cauchy approach was still in some ways dominant, for he called his study the "summation of divergent series," which sounds to be an uninteresting exercise if it is not realized that he meant "the nonorthodox summation of orthodoxly divergent series" (see Borel *1899a*, esp. ch. 1). He was slightly anticipated, in the special case of summability by arithmetic means, by E. Césaro (see Césaro *1890a*). The whole subject of the summation of "divergent" series is given a detailed treatment in Hardy *1949a*; and brief historical remarks may be found in Gloden *1950a*, pp. 212–220.

H. Burkhardt wrote an account *1911a* of the study of nonconvergent series during the period 1750–1860, without making clear that many of the issues arose from misunderstandings of the orthodox theory of convergence.

[31] See Sylow *1902a*, p. 57.

which he had seen to be short of full examination. But now he could bring to it the theory of convergence from Cauchy's *Cours d'analyse*, as well as incorporating a Cauchy-like extension of the problem to the case where m and x took complex values. "I dare say that it is the first perfectly rigorous proof of the binomial formula for all possible cases," he wrote to Holmboe at the end of the year, when the paper was in the press for Crelle's journal, "as well as of a large quantity of other partly known but weakly founded formulae."[32]

Abel wrote his papers for Crelle's journal in French, but with one exception his contributions to the first two volumes (including this one) were published in Crelle's translations from French into German. The original manuscripts did not survive, and so Holmboe translated the papers for his edition of Abel's work in 1839. The edition became of great rarity and the continuing interest in Abel's achievements led to the publication of a second edition by Ludwig Sylow (1832–1918) and Sophus Lie (1842–1899) in 1881. A fire at Holmboe's house had destroyed some of Abel's papers, but several more were discovered by Holmboe's widow, and in the new edition several Abelian juvenilia and Holmboe's interesting obituary were omitted, while additions included some late manuscripts, a few papers omitted by intention or oversight by Holmboe, and the 1826 paper on elliptic functions whose manuscript Holmboe had tried to obtain.

The "other formulae" of which Abel had written to Holmboe were the new theorems on convergence which he introduced in the first part of the paper to continue the achievements of Cauchy's *Cours d'analyse* "which," as he urged in its opening pages, "must be read by every analyst who likes rigor in mathematical researches."[33] Like Cauchy, Abel made no distinction between series of constant terms and series of functions and he also reasoned with infinitesimals. But he did not follow Cauchy's

[32] Abel *Letters*, p. 46; *Correspondence*, p. 54. Also in *Works*$_1$, vol. 2, p. 271; *Works*$_2$, vol. 2, p. 261. The letter is undated, but Holmboe assigns it to December, 1826. Earlier that year Abel published a short note in Crelle's journal on the expansion of $(x+\alpha)^n$ in powers of $(x+\beta)$, where $\alpha \neq \beta$ (Abel *1826a*). Doubtless Abel had seen Cauchy's discussion of the series in the *Cours d'Analyse* (Cauchy *1821a*, pp. 164–166; *Works*, ser. 2, vol. 3, pp. 146–147): in view of his interest in Bolzano (see n. 9 and text, p. 52) it is interesting to wonder if he knew Bolzano's pamphlet *1816a* on the series. Of the profound innovations of *1817a*, this work shows only the formulation of convergence in terms of the nth partial sum: otherwise it is merely a general study of various versions and manipulations of the series and so would have been of little use to Abel. The most notable section displays vestiges of limit-avoidance in considering the differentiation of a series term-by-term: in its course it uses the theorem of the title of *1817a*, and announces the preparation of that paper (see pp. 27–40, esp. 30 (footnote)).

[33] Abel *1826b*, p. 313; *Works*$_1$, vol. 1, p. 68; *Works*$_2$, vol. 1, p. 221.

general practice of distinguishing between a series and the series of its absolute values, and he never took the absolute values of the expressions he used in his proofs. Thus his analysis proves difficult to follow, and the question of what he had proved and what he thought he had proved grows in importance. Some of his new results were convergence tests, which we shall consider in the general context of such results; but there were other important theorems, such as "Abel's partial summation formula"

$$u_0 v_0 + u_1 v_1 + \cdots + u_n v_n =$$

$$(u_0 - u_1)s_0 + (u_1 - u_2)s_1 + \cdots + (u_{n-1} - u_n)s_{n-1} + u_n s_n, \tag{4.25}$$

where

$$s_r = \sum_{j=0}^{r-1} v_j, \tag{4.26}$$[34]

and "Abel's limit theorem," which he phrased as follows:

THEOREM 4.5
"If the series

$$f(\alpha) = v_0 + v_1\alpha + v_2\alpha^2 + \cdots + v_m\alpha^m + \cdots \tag{4.27}$$

is convergent for a certain value δ of α, it will also be convergent for every value less than δ, and, for always decreasing values of β, the function $f(\alpha - \beta)$ approaches indefinitely the limit $f(\alpha)$, presuming that α is equal to or smaller than δ."[35]

It is difficult to see immediately the significance of this theorem: in fact, the point is concealed in the word "equal" at the end, where we learn that if the power series is convergent at the boundary value δ of α, then it converges to the sum $f(\delta)$, which is therefore (Cauchy-) continuous at $\alpha = \delta$. Thus it is the first of those theorems of real-variable analysis which deal with the important question of *the behavior of a series of functions at the end-points of its range of convergence*. Take the case of the geometric progression

$$1 + x + x^2 + \cdots = \frac{1}{1-x}, \tag{4.28}$$

for example. One of the more mundane ways of "proving" that

$$1 - 1 + 1 - \cdots = \tfrac{1}{2}, \tag{4.29}$$

[34] Abel *1826b*, p. 314; *Works*$_1$, vol. 1, p. 69; *Works*$_2$, vol. 1, p. 222.
[35] Abel *1826b*, theorem IV. Abel wrote $f\alpha$ as opposed to our $f(\alpha)$.

as opposed to Euler's sophistications, was simply to put $x = -1$ in (4.28); but, as Bolzano pointed out in his 1817 paper and Cauchy echoed in the *Cours d'analyse*, (4.28) does *not* hold when $x = -1$.[36] And in 1826 Abel went further, to find a sufficient condition for the validity of such deductions in the prior convergence of the series so obtained.

This is the point of his theorem; but the proof is exceedingly difficult to follow, and seemingly suspect. Abel's aim was to show that $r_n(\alpha)$ is small for $\alpha < \delta$, and his previous results proved that

$$r_n(\alpha) < (\alpha/\delta)^n[\max_s(v_n\,\delta^n + \cdots + v_{n+s}\,\delta^{n+s})] \tag{4.30}$$

(it should be "$\leq$"); but their use required that $\alpha/\delta > 0$, which when added to the restriction $\alpha < \delta$ gives

$$0 < \alpha < \delta. \tag{4.31}$$

Abel also assumed a lower bound of zero on $r_n(\alpha)$, from which we then deduce

$$0 < \max_s(v_n\,\delta^n + \cdots + v_{n+s}\,\delta^{n+s}) \tag{4.32}$$

and thus

$$0 < \max_s(v_n + \cdots + v_{n+s}\,\delta^s), \tag{4.33}$$

since, from (4.31), $\delta > 0$. Now since (4.33) has to be true for all n it seems that all the v_n have to be positive. But this is an intolerable restriction on the original series (4.27) which Abel cannot have intended: one of its consequences, for example, is that (4.30) reduces in generality to

$$r_n(\alpha) < (\alpha/\delta)^n r_n(\delta). \tag{4.34}$$

The argument onwards from the smallness of $r_n(\alpha)$ for $\alpha < \delta$ is straightforward, but to reach that stage Abel's proof has been in great difficulties; and even if positive values are used, the difficulties are not all eliminated. The restriction (4.31) becomes

$$|\alpha| < |\delta|, \tag{4.35}$$

and (4.30) becomes

$$|r_n(\alpha)| \leq |\alpha/\delta|^n[\max_s(|v_n|\,|\delta|^n + \cdots + |v_{n+s}|\,|\delta|^{n+s})]. \tag{4.36}$$

[36] Bolzano *1817a*, art. 4. Cauchy *1821a*, pp. 124 and 126; *Works*, ser. 2, vol. 3, pp. 114–115 and 116.

The assumption of convergence of the series when $\alpha = \delta$ implies that $|r_n(\delta)|$ is small. But we cannot deduce the required smallness of $|r_n(\alpha)|$, for the only relevant analysis is to put $\alpha = \delta$ in (4.36) and obtain

$$|r_n(\delta)| \leq |v_n| \, |\delta|^n + \cdots \tag{4.37}$$

and the smallness of the left-hand side of (4.37) does not imply the smallness of the right-hand side, which we would then wish to use in the right-hand side of (4.36) to show the smallness of $|r_n(\alpha)|$.[37]

Abel's other main theorem was a generalization of the limit theorem to the case where the v_r are themselves continuous functions of another variable x. We may state the theorem in the following way:

THEOREM 4.6
If the functions $v_0(x)$, $v_1(x)$, ... are continuous (in Cauchy's sense) when $a \leq x \leq b$, and

$$v_0(x) + v_1(x)\,\delta + v_2(x)\,\delta^2 + \cdots \tag{4.38}$$

is convergent, then the series

$$f(x, \alpha) = v_0(x) + v_1(x)\alpha + v_2(x)\alpha^2 + \cdots, \tag{4.39}$$

where $\alpha < \delta$, is convergent and a continuous function of x when $a \leq x \leq b$.[38]

The proof is suspect in ways similar to its predecessor: indeed, Abel claimed that the convergence part of the theorem followed from it, and he used the same method to show the continuity of $f(x, \alpha)$ with respect to x. The most interesting aspect of the theorem is its relation to Cauchy's theorem 4.4 of the *Cours d'analyse*, on which Abel now inserted his footnote: "... it seems to me that that theorem admits of exceptions. For example, the series

$$\sin x - \tfrac{1}{2}\sin 2x + \tfrac{1}{3}\sin 3x - \cdots \tag{4.40}$$

[37] See Abel *1826b*, proof of theorem IV. Abel wrote $\phi\alpha$ for the nth partial sum and $\psi\alpha$ for the remainder term, and used m for our n. Sylow and Lie offered a spirited defense of Abel's proof (see Abel *Works*$_2$, vol. 2, p. 302): it would seem that they were not completely aware of the standards of analysis in their own time.

[38] Abel *1826b*, theorem V. Abel next applied his theorems to the binomial series. Other parts of his letter to Holmboe of January, 1826 on the horror of nonconvergent series (see n. 29 and text, p. 80) show that he had already seen the problem which his limit theorems were to solve. Sylow and Lie do not date the paper closer than to the winter period of 1825–26 (see Abel *Works*$_2$, vol. 2, p. 302). Crelle revealed his inability to appreciate Abel's work by publishing his own alleged proof of the binomial theorem based on the transformation of identities, shortly after Abel's paper (see Crelle *1829a* and *1830a*; and also the remarks in Mittag-Leffler *1927a*).

is discontinuous for all values $(2m+1)\pi$ of x, m being a whole number. There are, as one knows, many series of this type," from which we may presume that Fourier's book on heat diffusion was also in Crelle's library.[39]

Abel's theorem 4.6, like its predecessor, is directed at the behavior of the series (4.38) at the boundary value δ of α, and assumes the convergence with respect to x which was the subject of Cauchy's theorem 4.4. But boundary value problems with regard to x also occur with Fourier series, as Abel well knew. In his letter to Holmboe of January, 1826, he said of the example (4.40) of his later footnote that "it seems to follow that this formula should hold for $x=\pi$: but there one will find that

$$\frac{\pi}{2}=\sin\pi-\frac{1}{2}\sin 2\pi+\frac{1}{3}\sin 3\pi-\text{etc.}=0 \text{ (absurd),"}$$

which itself would refute Cauchy's theorem.[40]

So Abel was thinking about the convergence of Fourier series:[41] in fact, his generalized limit theorem challenged the only mathematical proof of convergence which had been published by 1826. One cannot be certain that Abel was aware of that proof, but without doubt Cauchy was acquainted with it, for he attacked it in the same year of 1826—and in his own inimitable way.

39 Abel *1826b*, footnote to theorem V.
The example cited by Abel is the Fourier sine series of $x/2$ over $0 \leq x \leq \pi/2$ and appears in Fourier's book (Fourier *1822a*, art. 222). It was first proved (as a special result) by Euler (see Euler *1754a*, p. 204; *Works*, ser. 1, vol. 14, p. 584).

40 Abel *Letters*, pp. 17–18; *Correspondence*, p. 18. Also in *Works*$_1$, vol. 2, p. 268; *Works*$_2$, vol. 2, p. 258.

41 He told Holmboe in his letter of December, 1826 that he hoped to write a paper on the convergence of Fourier series himself (Abel *Letters*, p. 47; *Correspondence*, p. 54). But his later manuscripts on convergence do not deal with it (see Abel *Series*, and commentary in Sylow *1902a*, pp. 51–57).

THE CONVERGENCE PROBLEM OF FOURIER SERIES

5

Partly in connection with the vibrating string problem, Daniel Bernoulli and Euler had a debate in the 1770s on trigonometric series and their properties.[1] Of the discussion Poisson wrote: "We admit with Euler that the sums of these series, considered in themselves, do not have determined values; but we would add that each of them has a unique value and that one can use them in analysis, when one regards them as limits of convergent series, that is to say, when one implicitly assumes their successive terms multiplied by powers of a fraction infinitely little different from unity."[2] In other words, a trigonometric series such as

$$a \cos x + b \cos 2x + c \cos 3x + \cdots \tag{5.1}$$

might lead to trouble because of divergence; but

$$pa \cos x + p^2 b \cos 2x + p^3 c \cos 3x + \cdots, \qquad 0 < p < 1, \tag{5.2}$$

will safely be convergent, and so (5.1) can be taken to be the limiting value of (5.2) when $p = 1$.

So Poisson, writing in 1823, did not understand what Abel appreciated in 1826; and nor did he after 1826, for he retained this staple component of his analysis throughout his life.[3] Already in 1820, he had used it to develop the first full mathematical proof of the convergence of Fourier series, in connection with his own work on heat diffusion; and, despite later developments, he made only minor variations of presentation. Since the basic approach assumes Abel's generalized limit theorem (theorem 4.6) and therefore the convergence of the Fourier series which is supposed to be being demonstrated, it is doomed from the start; but it is worth describing, both as a case study of the analysis of the time and also for the influence that it had on Poisson's successors.

Poisson started safely enough: summing $[1 + 2\sum_{r=1}^{\infty} p^r e^{ir(x-\alpha)}]$, $|p| < 1$, as a geometrical progression and equating the real parts of series and sum, he obtained a result of the type discussed by Euler and Bernoulli:

$$\frac{1 - p^2}{1 - 2p\cos(x - \alpha) + p^2} = 1 + 2\sum_{r=1}^{\infty} p^r \cos r(x - \alpha). \tag{5.3}$$

[1] See especially Bernoulli D. *1772a* and *1773a*, and Euler *1773a*.
[2] Poisson *1823c*, pp. 408–409: see also pp. 428–430.
[3] Poisson's views on summation were probably part of Cauchy's target in the introduction to the *Cours d'analyse* (see n. 2 and text, p. 49): "*a divergent series does not have a sum*," he put in italics (Cauchy *1821a*, introduction, p. iv).

Multiplying through by $f(\alpha)$ and integrating over $[-\pi, +\pi]$ of α gave him the "Poisson integral"

$$\int_{-\pi}^{+\pi} \frac{(1-p^2)f(\alpha)}{1-2p\cos(x-\alpha)+p^2}\,d\alpha = \int_{-\pi}^{+\pi} f(\alpha)\left[1+2\sum_{r=1}^{\infty} p^r \cos r(x-\alpha)\right] d\alpha \tag{5.4}$$

$$= G(p), \quad \text{say.} \tag{5.5}$$

As long as the integrals in (5.4) can be defined, the above reasoning is correct; but now Poisson took the fatal step of putting $p=1$ on both sides of (5.4). The left-hand side of (5.4) has an integrand of the form 0/0, and so needs careful handling. Poisson dealt with it in an unsatisfactory way: considering the component I_x of that integral corresponding to the interval $[x-\varepsilon, x+\varepsilon']$, where ε and ε' are small, he substituted

$$x-\alpha=z \quad \text{and} \quad p=1-g, \tag{5.6}$$

where g and z are small, and so deduced

$$\cos(x-\alpha)=\cos z = 1-\tfrac{1}{2}z^2. \tag{5.7}$$

But he also found that

$$f(\alpha)=f(x-z)=f(x), \tag{5.8}$$

which is infinitesimal reasoning at its best, and obtained in the left-hand side of (5.4) after all these substitutions

$$I_x = \int_{-\varepsilon}^{+\varepsilon'} \frac{[1-(1-g)^2]f(x)}{1-2(1-g)(1-\frac{1}{2}z^2)+(1-g)^2}\,dz$$

$$= f(x)\int_{-\varepsilon}^{+\varepsilon'} \frac{2g-g^2}{g^2+(1-g)z^2}\,dz. \tag{5.9}$$

He then allowed himself further suitable approximations to give

$$I_x = f(x)\int_{-\varepsilon}^{+\varepsilon'} \frac{2g}{g^2+z^2}\,dz \tag{5.10}$$

$$= 2f(x)\left[\tan^{-1}\left(\frac{\varepsilon'}{g}\right)+\tan^{-1}\left(\frac{\varepsilon}{g}\right)\right] \tag{5.11}$$

$$= 2\pi f(x), \tag{5.12}$$

where the limiting value 0 of g (that is, 1 of p) is taken in (5.11). Since the rest of the integral on the left-hand side of (5.4) is clearly zero, we obtain from (5.5) and (5.12),

$$G(1) = 2\pi f(x). \tag{5.13}$$

Putting $p = 1$ on the other side of (5.4) gives

$$G(1) = \int_{-\pi}^{+\pi} \left[1 + 2 \sum_{r=1}^{\infty} \cos r(x - \alpha)\right] d\alpha. \tag{5.14}$$

Therefore, equating (5.13) and (5.14), we obtain

$$f(x) = \frac{1}{2\pi} \int_{-\pi}^{+\pi} f(\alpha)\, d\alpha + \frac{1}{\pi} \sum_{r=1}^{\infty} \left[\cos rx \int_{-\pi}^{+\pi} f(\alpha) \cos r\alpha\, d\alpha + \sin rx \int_{-\pi}^{+\pi} f(\alpha) \sin r\alpha\, d\alpha\right], \tag{5.15}$$

which shows that the full Fourier series of $f(x)$ over $[-\pi, +\pi]$ does in fact converge to $f(x)$.[4]

Poisson's proof falters at virtually every step after (5.5), and for reasons additional to the assumption of convergence required by taking the limiting value of p there. The approximation (5.8) spoils (5.9), although (5.9) can be obtained by the use of Taylor's theorem for functions for which it is applicable: the further approximations on (5.10) are chosen with what might be politely described as "good taste," although once again, (5.11) is obtainable from (5.9) by more careful means; but then we have not only the use of the limiting value of p (or g), but also a switch of order in the double limit

$$\lim_{g \to 0} \lim_{\varepsilon, \varepsilon' \to 0} \int_{-\varepsilon}^{+\varepsilon'}$$

on (5.11) to deduce (5.12) and then an interchange of sum and integral in order to derive (5.15) from (5.14), both of which assume again the convergence of the series.[5]

Cauchy chose not to welcome the proof, and expressed his dissatisfaction in characteristic style. His silent remark on Fourier series in theorem

[4] Poisson's principal versions of this proof are, in chronological order: *1820a*, pp. 422–424; *1823b*, pp. 152–156; *1823c*, pp. 432–456 (including a study of finite trigonometric series and the Fourier integral theorem); and *1835a*, pp. 187–191. In *1823d* he used the method to evaluate various definite integrals. For descriptive commentary, see Paplauskas *1966a*, pp. 68–74.

[5] Poisson's proof-method has attracted some later writers. See, for example, Schwarz *1872a*, pp. 225–231; Bonnet *1879a*; Harnack *1888a*, pp. 195–202; and Bôcher *1906a*, pp. 92–98.

4.4 from the *Cours d'analyse* was not his last thought on the problem, for in February 1826 he presented his own proof of convergence. Firstly, however, he worked through a version of Poisson's proof for the purpose of then saying: "the preceding series can be very usefully employed in many circumstances. But it is important to show its convergence."[6] In other words, "this series is usefully employed by various unnamed persons. But the intuitive reasoning I have just used to derive it must not be mistaken for a genuine convergence proof."

Now he launched into his own proof. One of its faults may be mentioned at once, for the proof can be presented and indeed shortened without it. Cauchy used the full series over $[0,a]$ and so was anxious to establish that

$$f(x) = \frac{1}{a}\left\{\int_0^a f(t)\,dt + 2\sum_{r=1}^{\infty}\int_0^a f(t)\cos\left[\frac{2r\pi}{a}(x-t)\right]dt\right\}. \tag{5.16}$$

In his proof he operated on the whole series, but in our version we carry out the same analysis only on the general term of the series

$$v_n = \frac{2}{a}\int_0^a f(t)\cos\left[\frac{2n\pi}{a}(x-t)\right]dt \tag{5.17}$$

and so eliminate the lengthy and question-begging interchanges of sum and integral (notwithstanding his own theorem 3.5 from the *Résumé* on integrating a convergent series term-by-term) and rearrangement of order of terms which Cauchy used.

The argument is based on a theorem from his new theory of complex variable analysis:

THEOREM 5.1
If $f(t)$ is finite for all (real and complex) values of t and if $b > 0$, then

$$\int_0^a e^{\pm ibt} f(t)\,dt = \mp\int_0^{\infty}[e^{\pm iab}f(a \pm iu) - f(\pm iu)]e^{-bu}\,du. \tag{5.18}$$

If we write (5.17) as

$$v_n = \frac{1}{a}\int_0^a f(t)\exp\left[\frac{2n\pi i}{a}(x-t)\right]dt$$
$$+ \frac{1}{a}\int_0^a f(t)\exp\left[\frac{-2n\pi i}{a}(x-t)\right]dt \tag{5.19}$$

[6] The introductory section of the paper is in Cauchy *1826a*, pp. 603–606; *Works*, ser. 1, vol. 2, pp. 12–14.

and take

$$b = \frac{2n\pi}{a}, \tag{5.20}$$

then we can use theorem 5.1 with the lower choice of signs in (5.18) for the first integral on the right-hand side of (5.19), and the upper choice of signs for the second integral, and so obtain

$$\begin{aligned} v_n = {} & \frac{ie^{\frac{2n\pi ix}{a}}}{a} \int_0^a [e^{-2n\pi i} f(a - iu) - f(-iu)]\, e^{-\frac{2n\pi u}{a}}\, du \\ & - \frac{ie^{-\frac{2n\pi ix}{a}}}{a} \int_0^a [e^{2n\pi i} f(a + iu) - f(iu)]\, e^{-\frac{2n\pi u}{a}}\, du. \end{aligned} \tag{5.21}$$

Putting

$$z = \frac{2n\pi u}{a} \tag{5.22}$$

in (5.21) reduces it to

$$\begin{aligned} v_n = {} & \frac{ie^{\frac{2n\pi ix}{a}}}{2n\pi} \int_0^{2n\pi} \left[f\left(a - \frac{iaz}{2n\pi}\right) - f\left(-\frac{iaz}{2n\pi}\right) \right] e^{-z}\, dz \\ & - \frac{ie^{-\frac{2n\pi ix}{a}}}{2n\pi} \int_0^{2n\pi} \left[f\left(a + \frac{iaz}{2n\pi}\right) - f\left(\frac{iaz}{2n\pi}\right) \right] e^{-z}\, dz. \end{aligned} \tag{5.23}$$

As $n \to \infty$, each integrand in (5.23) tends to $[f(a) - f(0)]e^{-z}$, and so $v_n \to w_n$, where

$$\begin{aligned} w_n = {} & \frac{ie^{\frac{2n\pi ix}{a}}}{2n\pi} [f(a) - f(0)] \int_0^\infty e^{-z}\, dz \\ & - \frac{ie^{-\frac{2n\pi ix}{a}}}{2n\pi} [f(a) - f(0)] \int_0^\infty e^{-z}\, dz \\ = {} & -\frac{\sin \frac{2n\pi x}{a}}{n\pi} [f(a) - f(0)]. \end{aligned} \tag{5.24}$$

"Now it is clear that the series which has the expression $[w_n]$ for general term will be a convergent series," wrote Cauchy, and so completed the proof.[7]

The way that the proof does end is unclear. It is certainly true that $\Sigma_{r=1}^{\infty} w_r$ is convergent: in fact, by comparison of the expression for w_n in (5.24) with the Fourier series

$$\frac{\pi - x}{2} = \sin x + \frac{1}{2}\sin 2x + \frac{1}{3}\sin 3x + \cdots, \qquad 0 < x \leq \frac{\pi}{2}, \tag{5.25}$$

whose own convergence is demonstrable by other means, we can see that

$$\begin{aligned} \sum_{r=1}^{\infty} w_r &= -\left[\frac{f(a) - f(0)}{\pi}\right]\left[\frac{\pi - 2\pi x/a}{2}\right] \\ &= \frac{1}{2}[f(a) - f(0)]\left[\frac{2x}{a} - 1\right] \end{aligned} \tag{5.26}$$

over $(0, a/2]$ and the negative of this value over $[-a/2, 0)$, with a discontinuity at $x = 0$. But the relevance of $\Sigma_{r=1}^{\infty} w_r$ is difficult to understand. If Cauchy thought that $\Sigma_{r=1}^{\infty} v_r$ and $\Sigma_{r=1}^{\infty} w_r$ converged to the same sum, then, from (5.16), (5.17) and (5.26), he would have deduced that the Fourier series converged to the value

$$\frac{1}{a}\int_0^a f(t)\, dt \mp \frac{1}{2}[f(a) - f(0)]\left[\frac{2x}{a} - 1\right] \tag{5.27}$$

(where the choice of signs is governed by whether x is within $[-a/2, 0)$ or $(0, a/2]$), which bears not the slightest resemblance to $f(x)$. If, however, he thought that $\Sigma_{r=1}^{\infty} v_r$ and $\Sigma_{r=1}^{\infty} w_r$ summed to different values, then the question of the sum of the convergent Fourier series remained unresolved. Without doubt, he did assume that the convergence of $\Sigma_{r=1}^{\infty} w_r$

[7] Cauchy *1826a*, pp. 606–608; *Works*, ser. 1, vol. 2, pp. 14–16. Cauchy's equation (12) should end with the term $\sin \frac{2n\pi x}{a}$ and not $\sin \frac{2n\pi}{a}$, as has been printed both in the original and in the Collected Works. For descriptive commentary, see Paplauskas *1966a*, pp. 74–77.

In 1827 Cauchy offered another proof where the series itself is deduced from theorems of complex variable analysis. These theorems, however, seem to assume the convergence sought, and in any case severely limit the kinds of function for which convergence could be established (see Cauchy *1827c*, esp. pp. 356–358 and 363–365; *Works*, ser. 2, vol. 7, pp. 409–411 and 417–419). For discussion of this proof see Harnack *1888a*, pp. 175–195, and for descriptive commentary see Paplauskas *1966a*, pp. 78–81.

implied the convergence of $\Sigma_{r=1}^{\infty} v_r$; but even there he was in error, and so his proof of convergence was invalid. This was pointed out by a young man who understood the Fourier series problem for what it was—Peter Lejeune-Dirichlet (1805–1859).[8]

Like Abel and every aspiring young mathematician of the time, Dirichlet came to Paris in the 1820s. He formed a strong attachment to Fourier and worked on some problems suggested by Fourier series and heat diffusion. The main paper which resulted was a masterpiece of 1829 on the convergence of the Fourier series which he published in Crelle's journal: it can be seen as the product of a personal effort on behalf of the dying Fourier, for his main interest in mathematics lay elsewhere.[9]

Dirichlet decided to try and solve the convergence problem of Fourier series when he realized that the sum of a conditionally convergent infinite series could be changed by reordering its terms.[10] As he wrote in a later paper, a convergent series of positive terms took the same sum whatever order was used, but for a series of mixed terms if "convergence takes place for a certain order, it can stop through changing this order, or if this is not the case, the sum of the series can become quite different," and he cited the examples of the convergent

$$1-\frac{1}{\sqrt{2}}+\frac{1}{\sqrt{3}}-\frac{1}{\sqrt{4}}+\frac{1}{\sqrt{5}}-\frac{1}{\sqrt{6}}+\cdots \tag{5.28}$$

as opposed to the divergent

$$1+\frac{1}{\sqrt{3}}-\frac{1}{\sqrt{2}}+\frac{1}{\sqrt{5}}+\frac{1}{\sqrt{7}}-\frac{1}{\sqrt{4}}+\cdots, \tag{5.29}$$

and also

$$1-\frac{1}{2}+\frac{1}{3}-\frac{1}{4}+\frac{1}{5}-\frac{1}{6}+\cdots \tag{5.30}$$

against

$$1+\frac{1}{3}-\frac{1}{2}+\frac{1}{5}+\frac{1}{7}-\frac{1}{6}+\cdots, \tag{5.31}$$

which converge to different values.[11]

[8] For information on Dirichlet's life see the obituary Kummer *1860a*, and the documents in Biermann *1959a*.

[9] The paper is Dirichlet *1829a*. A paper of 1830 developed the solution to a heat diffusion problem which Fourier had given without proof (see Dirichlet *1830a*; and Fourier *1829a*, arts. 1–10).

[10] According to a report by Riemann (see Riemann *1866a*, art. 3).

[11] Dirichlet *1837d*, pp. 48–49; *Works*, vol. 1, pp. 318–319.

The discovery was important for analysis, for it ushered in the problem of *rearrangement of terms in a series* (which had usually been taken for granted), and also highlighted the distinction between series of terms with the same and with mixed signs. It was important for Dirichlet, too, because it heightened the problem of Fourier series in an interesting way. For this problem is *more than a convergence problem*: it is a summation problem as well, for the task is to show not only that the series—with its coefficients *defined* by the integral forms—is convergent, but also that it converges to the function. It is this combination of the two ideas, which Cauchy does not seem to have appreciated, which lends it its especial difficulty.

Dirichlet opened his 1829 paper with a review of the "state of the art"—such as it was. In fact, after a brief allusion to the original paper of 1807 which Fourier must have shown him during his Paris visit, he mentioned Cauchy's 1826 paper as the only one on the convergence problem of which he was aware. "The author of this work himself acknowledges that his proof is defective for certain functions for which, however, convergence is incontestable," he wrote,[12] presumably quoting a remark made by Cauchy (off-guard!). Dirichlet made a variety of criticisms of Cauchy's paper: most of them, being directed against the use of complex numbers, would appear to be misdirected since the use of such methods seems valid within the limitations placed on the function. But then he sounded the death knell of the proof by showing the falsity of Cauchy's assumption that the convergence of $\Sigma_{r=1}^{\infty} w_r$ implies the convergence of $\Sigma_{r=1}^{\infty} v_r$ with the counterexample:

$$w_n = \frac{(-1)^n}{\sqrt{n}}, \text{ for which } \sum_{r=1}^{\infty} w_r \text{ is convergent,} \tag{5.32}$$

and

$$v_n = \frac{(-1)^n}{\sqrt{n}} + \frac{1}{n}, \text{ for which } \sum_{r=1}^{\infty} v_r \text{ is divergent,} \tag{5.33}$$

although $v_n \to w_n$ as $n \to \infty$, as in Cauchy's proof.[13]

Dirichlet took his own approach to the problem from Fourier's book on heat diffusion. After developing the theory of the "Fourier integral theorem," which we write here in the form

$$f(x) = \frac{1}{\pi} \int_{-\infty}^{+\infty} f(\alpha)\, d\alpha \int_0^{\infty} \cos p(x - \alpha)\, dp, \tag{5.34}$$

[12] Dirichlet *1829a*, p. 157; *Works*, vol. 1, p. 119.
[13] Dirichlet *1829a*, p. 158; *Works*, vol. 1, p. 120.

Fourier had given himself over to a general consideration of the mathematics that he had used in the book.[14] At one stage he turned (5.34) into

$$f(x) = \frac{1}{\pi} \int_{-\infty}^{+\infty} f(\alpha) \frac{\sin p(x-\alpha)}{x-\alpha} d\alpha \tag{5.35}$$

with p infinite, and discussed its geometrical interpretation as a finite sequence of components of alternating sign created by the oscillation of $\sin p(x-\alpha)$.[15] Shortly afterwards he outlined a similar proof for the convergence of his series. There were several minor mistakes in the argument, but it may be put as follows. We take the full Fourier series over $[-X/2, +X/2]$:

$$\begin{aligned} f(x) &= \frac{1}{X} \int_{-X/2}^{+X/2} f(\alpha)\, d\alpha + \frac{2}{X} \sum_{r=1}^{\infty} \left[\cos \frac{2r\pi x}{X} \int_{-X/2}^{+X/2} f(\alpha) \cos \frac{2r\pi\alpha}{X} d\alpha \right. \\ &\qquad \left. + \sin \frac{2r\pi x}{X} \int_{-X/2}^{+X/2} f(\alpha) \sin \frac{2r\pi\alpha}{X} d\alpha \right] \\ &= \frac{1}{X} \int_{-X/2}^{+X/2} f(\alpha) \sum_{r=-\infty}^{+\infty} \cos \left[\frac{2r\pi}{X} (\alpha - x) \right] d\alpha \end{aligned} \tag{5.36}$$

and use the equation

$$\sum_{r=-n}^{+n} \cos ru = \cos nu + \sin nu \frac{\sin u}{1-\cos u} \tag{5.37}$$

corresponding to the middle $(2n+1)$ terms of the summation in (5.36). We then put

$$u = \frac{2\pi}{X} (\alpha - x) \tag{5.38}$$

in (5.37), and the integral of (5.36) for those $(2n+1)$ terms is

$$\frac{1}{X} \int_{-X/2}^{+X/2} f(\alpha) \cos nu\, d\alpha + \frac{1}{X} \int_{-X/2}^{+X/2} f(\alpha) \sin nu \frac{\sin u}{1-\cos u} d\alpha. \tag{5.39}$$

Now the oscillation of $\cos nu$ shows that the first integral in (5.39) becomes zero when n tends to infinity. The same is true for $\sin nu$, but the behavior of the second integral in (5.39) is affected also by the

[14] Fourier *1822a*, esp. arts. 342–362 and 396–423.
[15] Fourier *1822a*, arts. 415 and 416.

factor $\frac{\sin u}{1-\cos u}$: "one realizes clearly," wrote Fourier, "that the integral $\int f(\alpha) \sin nu \frac{\sin u}{1-\cos u} d\alpha$ has subsisting values only for certain infinitely small intervals: that is to say, when the ordinate $\frac{\sin u}{1-\cos u}$ becomes infinite. That will happen if u or $\frac{2\pi}{X}(\alpha - x)$ is nothing: and, in the interval where α differs infinitely little from x, the value of $f(\alpha)$ is identical with $f(x)$."[16] Therefore the limiting value of the second integral of (5.39) is, from (5.38) and the approximations,

$$\lim_{n\to\infty}\left[\frac{2}{X}\int_0^\delta f(x)\sin nu\,\frac{u}{\frac{1}{2}u^2}\,\frac{X\,du}{2\pi}\right], \tag{5.40}$$

$$=\lim_{n\to\infty}\frac{2}{\pi}f(x)\int_0^\delta \frac{\sin nu}{u}\,du=f(x), \qquad \delta>0,$$

which establishes the convergence required.[17]

So, like Poisson and Cauchy, Fourier held a vague belief that convergence took place for "any" function; but he surpassed them both in the quality of his attack on the problem and his sketch of the argument provided Dirichlet with the start that he needed. Using the interval $[-\pi, +\pi]$ rather than Fourier's $[-X/2, +X/2]$ he took the alternative version of Fourier's (5.37):

$$\sum_{r=-n}^{+n} \cos ru = \frac{\sin(n+\frac{1}{2})u}{2\sin\frac{1}{2}u}$$

to produce the famous "Dirichlet partial sum formula" corresponding to Fourier's (5.39):

[16] Fourier *1822a*, art. 423. We have modified the symbolism to conform with our textual conventions.

[17] Fourier's mistakes in his book *1822a* are as follows:

1. In writing a version of the full Fourier series over $[0, X]$ (corresponding to our (5.36)) he prefaced the expression with $1/2\pi$ instead of $1/X$ (see art. 418, equation (A)). He then remarked that the series could be converted into the integral theorem (5.34) by taking an infinite interval and forming a second integral out of the summation over r. But then his mistake becomes nontrivial, for the missing $1/X$ is needed to form the second differential.
2. In quoting that series in art. 423, he still wrote $1/2\pi$, but took the interval of integration to be $[-X, +X]$ instead of $[0, X]$. Then, to "simplify expressions," he put $X=2\pi$!
3. In his equivalent of our (5.40) in art. 423, he used $[0, \infty]$ instead of $[0, \delta]$.

All these mistakes were faithfully preserved in the English translation of the book, but in his edition of it for Fourier's *Works*, vol. 1, Darboux silently corrected mistake 1, preserved the correction in 2 and converted the interval of integration into $[-X/2, +X/2]$, but did not modify 3.

$$s_n(x) = \frac{1}{\pi}\int_{-\pi}^{+\pi} f(t)\,\frac{\sin[(n+\frac{1}{2})(x-t)]}{2\sin[\frac{1}{2}(x-t)]}\,dt, \tag{5.41}$$

and then tackled the convergence problem in the new way of examining the behavior of $s_n(x)$ as n increased. Breaking up $[-\pi, +\pi]$ into the respective subintervals $[-\pi, x]$ and $[x, +\pi]$ and substituting

$$x - 2u = t \quad \text{and} \quad x + 2u = t \tag{5.42}$$

into the respective subintegrals of (5.41), he found that

$$\begin{aligned} s_n(x) = {} & \frac{1}{\pi}\int_0^{(\pi+x)/2} f(x-2u)\,\frac{\sin(2n+1)u}{\sin u}\,du \\ & + \frac{1}{\pi}\int_0^{(\pi-x)/2} f(x+2u)\,\frac{\sin(2n+1)u}{\sin u}\,du. \end{aligned} \tag{5.43}$$

Both integrals are examples of the general form

$$I = \int_0^h f(x)\,\frac{\sin mx}{\sin x}\,dx, \qquad h \leq \frac{\pi}{2}, \tag{5.44}$$

which is known as the "Dirichlet integral" because of the analysis made of it by him in this paper.[18]

As Fourier had observed, the crucial component of the integrand is $\sin mx$, which changes sign for each subinterval $[(j-1)\pi/m, j\pi/m]$ of $[0, h]$. If we define r to be the integer for which

$$\frac{r}{m}\pi < h < \frac{r+1}{m}\pi, \tag{5.45}$$

then there will be r subareas, plus a residue over $[r\pi/m, h]$, through which $\sin mx$ takes alternating signs. At the same time, since $h < \pi/2$, the denominator $\sin x$ increases monotonically. If we now let m tend to infinity—corresponding to the increase of n in $s_n(x)$—then these subareas will become infinite in number and so correspond to terms of an infinite series. If conditions imposed on $f(x)$ make this series convergent, then the integral, and therefore the Fourier series, will be convergent also. On the strength of Fourier's own reasoning the sum should be the function $f(x)$ itself.

[18] Dirichlet *1829a*, pp. 166–167; *Works*, vol. 1, pp. 128–129. In his paper he formed the nth partial sum *after* his analysis of the "Dirichlet integral."

This is the Fourier-inspired background to Dirichlet's thinking, and in order to obtain convergence of his integral he imposed on $f(x)$ the conditions of being continuous, positive and monotonically decreasing, so that the subareas would form a series of terms of alternating sign and decreasing magnitude. So convergence would seem to be well assured; but the argument would need to be backed by careful analysis, which would also have to ensure that the sum of the convergent series was in fact the function itself.

Using (5.45) in (5.44) Dirichlet found that

$$I = \sum_{j=1}^{r} \int_{(j-1)\pi/m}^{j\pi/m} f(x) \frac{\sin mx}{\sin x}\, dx + \int_{r\pi/m}^{\pi} f(x) \frac{\sin mx}{\sin x}\, dx \tag{5.46}$$

$$= \sum_{j=1}^{r} I_j + H, \quad \text{say.} \tag{5.47}$$

By the mean value theorem (theorem 3.3)

$$I_j = \rho_j \int_{(j-1)\pi/m}^{j\pi/m} \frac{\sin mx}{\sin x}\, dx \tag{5.48}$$

where

$$f\left((j-1)\frac{\pi}{m}\right) > \rho_j > f\left(j\frac{\pi}{m}\right), \tag{5.49}$$

and so

$$I_j = (-1)^{j+1} \rho_j K_j, \tag{5.50}$$

where K_j is the numerical value of the integral in (5.48). The substitution

$$y = mx \tag{5.51}$$

converts (5.48) into

$$I_j = \rho_j \int_{(j-1)\pi}^{j\pi} \frac{\sin y}{m \sin(y/m)}\, dy, \tag{5.52}$$

and when m is large

$$\sin\frac{y}{m} = \frac{y}{m}. \tag{5.53}$$

Hence in (5.52),

$$I_j = \rho_j \int_{(j-1)\pi}^{j\pi} \frac{\sin y}{y}\, dy \tag{5.54}$$

$$= (-1)^{j+1} \rho_j k_j, \tag{5.55}$$

where k_j is the numerical value of the integral in (5.54). Now a known result is that

$$\frac{\pi}{2} = \int_0^{\infty} \frac{\sin y}{y}\, dy = \sum_{j=1}^{\infty} \int_{(j-1)\pi}^{j\pi} \frac{\sin y}{y}\, dy, \tag{5.56}$$

which becomes, from (5.54) and (5.55),

$$\frac{\pi}{2} = \sum_{j=1}^{\infty} (-1)^{j+1} k_j. \tag{5.57}$$

Now, using (5.50) in (5.47), we find that

$$I = \sum_{j=1}^{r} (-1)^{j+1} k_j \rho_j + H, \tag{5.58}$$

and the task is to prove the convergence of I as m (and therefore r) increases, from the known convergence in (5.57). To this end, we write

$$I = (K_1 \rho_1 - K_2 \rho_2 + \cdots - K_m \rho_m) + (K_{m+1} \rho_{m+1} - \cdots) + H \tag{5.59}$$

$$= s_m + r_m \tag{5.60}$$

(m even), and note that (5.49) shows that each $\rho_j \to f(0)$ as $m \to \infty$, and therefore, from (5.50) and (5.55), that

$$K_j \to k_j. \tag{5.61}$$

Hence in (5.60),

$$0 < r_m \to (k_{m+1} - \cdots) f(0) + 0 \tag{5.62}$$

$$< k_{m+1} f(0), \tag{5.63}$$

since (5.54) shows that the $1/y$ in the integral diminishes the integrand as y increases, and therefore that the k_r successively decrease in value. In addition, the convergence in (5.57) implies that

$$k_{m+1} \to 0 \quad \text{as} \quad m \to \infty. \tag{5.64}$$

Therefore, from (5.62) and (5.63),

$$r_m \to 0 \quad \text{as} \quad m \to \infty \tag{5.65}$$

also, and so I is convergent as $m \to \infty$.

To what value does it converge? From (5.60) and (5.61),

$$s_m \to f(0) \sum_{r=1}^{m} (-1)^{j+1} k_j \quad \text{as} \quad m \to \infty, \tag{5.66}$$

and therefore, from (5.57) and (5.44), we have finally

$$\lim_{m\to\infty} [I] = \lim_{m\to\infty} \int_0^h f(x) \frac{\sin mx}{\sin x}\, dx = \frac{\pi}{2} f(0). \tag{5.67}$$

Dirichlet now made a number of extensions to this evaluation of I with regard to the properties of the function. Firstly, the analysis holds if $f(x)$ is monotonically increasing or negative over $[0, h]$—for then we use $[-f(x)]$ instead of $f(x)$—or if $f(x)$ is constant over the interval, since, if it takes the value C, (5.44) shows that the limiting value of I is

$$\lim_{m\to\infty} \int_0^h C \frac{\sin mx}{\sin x}\, dx$$

$$= \lim_{m\to\infty} \int_0^{mh} C \frac{\sin y}{\sin (y/m)} \frac{dy}{m}$$

from the substitution (5.51),

$$= \int_0^{\infty} C \frac{\sin y}{y}\, dy,$$

from (5.53), and

$$= C\frac{\pi}{2}, \tag{5.68}$$

from (5.57). Therefore the evaluation of I is true if $f(x)$ changes sign within $[0, h]$; for in such a case we would carry out the evaluation on the function $(C + f(x))$, where C is a constant large enough to ensure that the function was positive over $[0, h]$, and then subtract from it the same result for the constant function C. Next, the analysis would also be true over the interval $[0, g]$, where $0 < g < h$, and so by subtracting (5.67) from itself applied over $[0, g]$ we obtain

$$\lim_{m\to\infty} \int_g^h f(x) \frac{\sin mx}{\sin x}\, dx = 0. \tag{5.69}$$

Finally, if the function is discontinuous at $x=g$ and/or $x=h$, the analysis has to be modified so that $f(g+\varepsilon)$ is written for $f(g)$ and $f(h-\varepsilon)$ for $f(h)$, where ε is a small positive quantity. This is a most significant extension, as we shall see; but for now we gather together the basic result (5.67) and all its extensions into the following theorem:

THEOREM 5.2

Let $0<g<h\leq\pi/2$, and let $f(x)$ be continuous and monotonic over (g, h) (and, possibly, also at $x=g$ and $x=h$). Then

$$\lim_{m\to\infty}\int_g^h f(x)\frac{\sin mx}{\sin x}\,dx=\begin{cases}\dfrac{\pi}{2}f(\varepsilon) & \text{if } g=0\\[2ex] 0 & \text{if } g\neq 0.\end{cases} \tag{5.70}$$

[19]

Dirichlet now applied his theorem to the evaluation of the two integrals in his expression (5.43) for the nth partial sum of the Fourier series. We describe his analysis on the second integral

$$\frac{1}{\pi}\int_0^{(\pi-x)/2} f(x+2u)\frac{\sin(2n+1)u}{\sin u}\,du; \tag{5.71}$$

the first integral is handled in a similar manner.

Two cases have to be considered, depending on the value of x. If $0\leq x<\pi$, then the range of integration $[0, \frac{1}{2}(\pi-x)]$ is within $[0, \pi/2]$ and so theorem 5.2 can be used directly. We assume that the function has finite discontinuities or takes a turning value at the points d_1, d_2, d_3, ... d_s of $[0, \frac{1}{2}(\pi-x)]$ and is otherwise continuous. Then $f(x)$ satisfies the conditions of theorem 5.2 over each subinterval $[0, d_1)$, (d_1, d_2), (d_2, d_3), ..., $(d_s, \frac{1}{2}(\pi-x)]$, and so we learn that

$$\begin{aligned}&\frac{1}{\pi}\lim_{n\to\infty}\left[\int_0^{d_1}+\int_{d_1}^{d_2}+\cdots+\int_{d_s}^{(\pi-x)/2}\right]\left[f(x+2u)\frac{\sin(2n+1)u}{\sin u}\,du\right]\\ &=\frac{1}{\pi}\left[\frac{\pi}{2}f(x+\varepsilon)+0+\cdots+0\right]\\ &=\tfrac{1}{2}f(x+\varepsilon).\end{aligned} \tag{5.72}$$

The second case corresponds to $-\pi<x<0$, and here the range of integration has to be split into the components $[0, \frac{1}{2}\pi]$ and $[\frac{1}{2}\pi, \frac{1}{2}(\pi-x)]$. The latter component lies outside $[0, \frac{1}{2}\pi]$ but can be brought within its scope by the transformation

[19] Dirichlet *1829a*, pp. 159–165; *Works*, vol. 1, pp. 121–127.

$$u = \pi - v. \tag{5.73}$$

When this is done, (5.71) is converted to

$$\frac{1}{\pi}\int_0^{\pi/2} f(x+2u)\frac{\sin(2n+1)u}{\sin u}\,du + \frac{1}{\pi}\int_{(\pi+x)/2}^{\pi/2} f(x+2\pi-2v)\frac{\sin(2n+1)v}{\sin v}\,dv, \tag{5.74}$$

and theorem 5.2 shows that its limiting value is

$$\frac{1}{\pi}\left[\frac{\pi}{2}f(x+\varepsilon)+0\right]$$
$$= \tfrac{1}{2}f(x+\varepsilon), \tag{5.75}$$

as in the first case.

Two particular values of x remain: $-\pi$ and $+\pi$. When $x=-\pi$, then the interval has to be split into two components in the manner used to obtain (5.74); but now the second integral is over $[0, \tfrac{1}{2}\pi]$ of v, and so the limiting value becomes

$$\frac{1}{\pi}\left[\frac{\pi}{2}f(-\pi+\varepsilon)+\frac{\pi}{2}f(\pi-\varepsilon)\right]$$
$$= \tfrac{1}{2}[f(-\pi+\varepsilon)+f(\pi-\varepsilon)]. \tag{5.76}$$

When $x=\pi$ the basic interval becomes $[0,0]$ and therefore the integral is zero anyway.

This completes the examination of the second integral (5.71) of the nth partial sum (5.43). When the first has been treated similarly, we obtain a collection of results which may be presented in the following form:

Value of x	Limiting value of the first integral of the nth partial sum	Limiting value of the second integral of the nth partial sum	Limiting value of the nth partial sum
$-\pi$	0	$\tfrac{1}{2}[f(-\pi+\varepsilon)+f(\pi-\varepsilon)]$	$\tfrac{1}{2}[f(-\pi+\varepsilon)+f(\pi-\varepsilon)]$
$-\pi<x<+\pi$	$\tfrac{1}{2}f(x-\varepsilon)$	$\tfrac{1}{2}f(x+\varepsilon)$	$\tfrac{1}{2}[f(x-\varepsilon)+f(x+\varepsilon)]$
$+\pi$	$\tfrac{1}{2}[f(\pi-\varepsilon)+f(-\pi+\varepsilon)]$	0	$\tfrac{1}{2}[f(\pi-\varepsilon)+f(-\pi+\varepsilon)]$

Thus Dirichlet had proved

THEOREM 5.3
If $f(x)$ satisfies the "Dirichlet conditions" over $[-\pi, +\pi]$ of

1. being continuous and bounded, except perhaps for a finite number of finite discontinuities, and
2. possessing a finite number of turning values,

then its Fourier series converges to the value $\frac{1}{2}[f(x-\varepsilon)+f(x+\varepsilon)]$ over $(-\pi, +\pi)$, and, if we follow Fourier in interpreting $f(x)$ as being geometrically periodic outside $[-\pi, +\pi]$ (when $f(\pm\pi)=f(\mp\pi)$), over $[-\pi, \pi]$.[20]

So ends one of the most important proofs in the new era of Bolzano's "pure analysis." Dirichlet had formulated for the first time the convergence problem of Fourier series, and offered a solution to it which his successors could try to extend. He had shown that the problem was a question of finding sufficient conditions on a function for which its Fourier series converged to it, and that, under his conditions, the value of the sum at a point was the arithmetic mean of the left- and right-hand values of the function at that point. In this way he also introduced these values into the theory of functions, and gave some information on a question left untouched by Fourier: the value of the series at a point of discontinuity of the function, when such values are not equal. Dirichlet's understanding of the theory of functions was extraordinarily acute for his day, and it is from him more than anyone else that we have learned how to handle them. Perhaps the best example of his understanding comes from the end of this paper, when he gave an example of a function *not* satisfying his conditions:

"$f(x)$ equals a determined constant c when the variable x takes a rational value, and equals another constant d, when this variable is irrational. The function thus defined has finite and determined values for every value of x, and meanwhile one cannot substitute it into the series, seeing that the different integrals which enter into that series lose all significance in this case."[21]

The contemporary description of this function is "the characteristic function of the rationals," and both it and the "significance" of its

[20] Dirichlet *1829a*, pp. 166–169; *Works*, vol. 1, pp. 127–131. For a description of his proof, see Paplauskas *1966a*, pp. 85–92. It also has attracted later writers: see especially Stäckel *1901b*; Bôcher *1906a*, pp. 141–145; and Neumann *1914a*.
[21] Dirichlet *1829a*, p. 169; *Works*, vol. 1, p. 132.

integral assumed great importance in the later stages of mathematical analysis. It would be unfair to imagine that Dirichlet perceived their implications in 1829, but one would dearly wish to know what he had in mind for the "fundamental principles of infinitesimal analysis" or the "some other quite remarkable properties" of Fourier series which he promised at the close of his paper for future work.[22] For, his later writings on Fourier series made only marginal additions to what had already been achieved. A paper of 1837, written for the first volume of a new journal

[22] Dirichlet *1829a*, p. 169; *Works*, vol. 1, p. 132. The next paper in Crelle's journal, by E. H. Dirksen, was, like Dirichlet's, concerned with the convergence of Fourier series from consideration of the integral form of the partial sum, and was also dated "Berlin, January, 1829." But there was seemingly no relation to Fourier's work: even the series was cited from Euler (Dirksen *1829a*, p. 170; the reference was to Euler *1777a*, art. 3, as given also in n. 44, p. 19). The most relevant part of the analysis —longwinded and obscure after Dirichlet's lucidity—is based on integrating by parts the same kind of expression

$$I(\mu_0, \mu_1) = \int_{\mu_0}^{\mu_1} \frac{\sin[\pi(n+\frac{1}{2})(x-\mu)/\alpha]}{\sin[\pi(x-\mu)/2\alpha]} f(\mu)\, d\mu, \tag{1}$$

where the integrand corresponds to an interval of width 2α, to give

$$I(\mu_0, \mu_1) = \frac{1}{n+\frac{1}{2}} \sum_{r=1}^{m} \frac{\cos[\pi(n+\frac{1}{2})(x-\mu)/\alpha]}{\sin[\pi(x-\mu)/2\alpha]} [f(d_r-\varepsilon)-f(d_{r-1}+\varepsilon)]$$
$$+\frac{1}{n+\frac{1}{2}} \int_{\mu_0}^{\mu_1} \frac{\cos[\pi(n+\frac{1}{2})(x-\mu)/\alpha)]}{\sin[\pi(x-\mu)/2\alpha]} f'(\mu)\, d\mu, \tag{2}$$

where $d_0 = \mu_0$, $d_m = \mu_1$, and $d_1, \ldots, d_{m-1}$ are the points of discontinuity of $f(x)$ which is otherwise continuous and finite. But the denominator $\sin[\pi(x-\mu)/2\alpha]$ is seemingly allowed to be untouched by this process. Dirksen then argued that if $f'(\mu)$ is finite, (2) shows that $I(\mu_0, \mu_1) \to 0$ as $n \to \infty$ unless the chaste denominator is zero. This happens when

$$\mu = \pm 2N\alpha + x, \tag{3}$$

and a singular integral technique shows that the limiting value for each point is

$$\frac{1}{2\alpha}[2\alpha f(x \mp 2N\alpha)]. \tag{4}$$

For the particular interval $[-\alpha, +\alpha]$ the only such point is $\mu = x$, and therefore, from (1) and (4),

$$\lim_{n\to\infty} [I(-\alpha, +\alpha)] = f(x), \tag{5}$$

which proves the convergence required. But (2) would still seem to be problematic! (Dirksen *1829a*, esp. pp. 174–178. For commentary, see Paplauskas *1966a*, pp. 81–84.) Dirksen may have written the paper on hearing of Dirichlet's work, for it was a development of the even more obscure paper *1827a* sent to the Berlin Academy, where Fourier *was* mentioned. Just after completing his paper for Crelle he sent another paper to the Berlin Academy in which the sine denominator of (2) *was* handled, by means of

$$\sin z = z - \frac{z^2}{2!} \sin \lambda z, \qquad 0 < \lambda < 1. \tag{6}$$

(See Dirksen *1829b*, especially the theorem in equation (35).)

with which he was associated, was basically a repetition of its predecessor, and became such a "classic" presentation of his method that it eclipsed the masterpiece of his 24th year.[23] Its chief points of interest are a series of interesting arguments for the plausibility of Fourier series, the emphasis on the need for the functions to be single-valued, and the introduction of the now standard notations $f(x-0)$ and $f(x+0)$ for the left- and right-hand values of a function at a point; on the other hand, the strange counter-example was omitted.[24] But to another paper of 1837 he attached an appendix extending his conditions for convergence to the case where $f(x)$ took infinite values by means of Cauchy's "singular integrals." If $f(x)$

[23] Dirichlet *1837a*: the journal was the *Reportorium der Physik*. In a slightly earlier paper he applied his results to the summation of various trigonometric series (see *1835a* and the extract *1837c*), and in 1853 he wrote to Gauss about his integral (see Dirichlet *Works*, vol. 2, pp. 385–387).

[24] See Dirichlet *1837a*, pp. 152–160, 152 and 170 respectively; *Works*, vol. 1, pp. 135–145, 135 and 156. The stress on the need for single-valued functions, and also the use of continuous and discontinuous functions, may be traced to the same part of Fourier's book which contains the genesis of Dirichlet's proof-method (see Fourier *1822a*, art. 417).

The most interesting of Dirichlet's plausibility arguments for Fourier series also derived from Fourier. Fourier had tried unsuccessfully to achieve the analysis of heat diffusion in continuous bodies by solving the problem for an n-body discrete model and then letting n tend to infinity (see n. 48, p. 20). In the course of his attempt he had to solve for $a_1, \ldots, a_{n-1}$ the equations

$$\sum_{r=1}^{n-1} a_r \sin rx = f(x), \qquad x = \frac{k}{n}\pi, \qquad k = 1, \ldots, (n-1). \tag{1}$$

His solution was

$$a_m = \frac{2}{\pi}\sum_{r=1}^{n-1} f\left(r\frac{\pi}{n}\right)\sin\left(mr\frac{\pi}{n}\right)\frac{\pi}{n}, \qquad m = 1, \ldots, (n-1) \tag{2}$$

(Fourier *1807a*, art. 10; *1811a*, art. 40; *1822a*, arts. 265–271). Dirichlet now pointed out that if we let n tend to infinity then the expression under summation in (2) could be taken as a sum, such as used in forming the integral of $f(x)\sin mx$ via a regular partition of $[0, \pi]$, and so its limiting value when n is taken to infinity is

$$a_m = \frac{2}{\pi}\int_0^\pi f(x)\sin mx\,dx, \tag{3}$$

which is precisely the Fourier coefficient for the corresponding sine series (Dirichlet *1837a*, pp. 156–158; *Works*, vol. 1, pp. 140–142). In fact he could have gone further with this kind of reasoning to multiply (1) through by $\sin mx$ as it stands and integrate over $[0, \pi]$ to give exactly (3). Thus, if $f(x)$ is identified with the polynomial passing through the n points in (1) we have the solution

$$a_m = \frac{2}{\pi}\sum_{r=1}^{n-1} f\left(r\frac{\pi}{n}\right)\sin\left(mr\frac{\pi}{n}\right)\frac{\pi}{n} = \frac{2}{\pi}\int_0^\pi f(x)\sin mx\,dx, \tag{4}$$

and the process of taking n to infinity could be seen as extending the number of points specified for $f(x)$. Yet this reasoning was not suggested for more than a century after Dirichlet's previous idea, presumably because other developments overtook any use that could be made of it (see Thwaites *1960a*, pp. 119–120; and *1967a*, p. 143).

is infinite at $x = c$ within $[0, h]$, then the Dirichlet integral (5.44) must be broken up into four components:

$$\left[\int_0^{c-\varepsilon} + \int_{c-\varepsilon}^{c} + \int_c^{c+\varepsilon} + \int_{c+\varepsilon}^{h}\right]\left[f(x)\frac{\sin mx}{\operatorname{tin} x}\,dx\right]$$

$$= I_1 + I_2 + I_3 + I_4, \quad \text{say.} \tag{5.77}$$

I_1 and I_4 are Dirichlet integrals in the sense of his theorem 5.2, and so their limiting values are $(\pi/2)f(\varepsilon)$ and 0 respectively. I_2 and I_3, however, are singular integrals, and for them Dirichlet chose ε small enough for $f(x)$ to keep the same sign in $(c - \varepsilon, c)$ and in $(c, c + \varepsilon)$. (Of course, there may be a change of sign from one interval to the other.) If we write $F(x)$ for the indefinite integral of $f(x)$, then, from (5.77),

$$|I_2| < \frac{1}{\sin(c - \varepsilon)}\left|\int_{c-\varepsilon}^{c} f(x)\,dx\right|$$

or

$$|I_2| < \frac{1}{\sin(c - \varepsilon)}|F(c) - F(c - \varepsilon)|. \tag{5.78}$$

Similarly,

$$|I_3| < \frac{1}{\sin c}|F(c + \varepsilon) - F(c)|. \tag{5.79}$$

Now $F(x)$ is continuous: so the numerators of both (5.78) and (5.79) are small, and therefore I_2 and I_3 tend to zero. Hence the limiting value of the Dirichlet integral, under these extended conditions on the function $f(x)$, is still $(\pi/2)f(\varepsilon)$. The application of theorem 5.2 to theorem 5.3 proceeds as before, and so the Fourier series is convergent to a function which also possesses a finite number of infinities.[25]

The main part of this second paper of 1837 indicates another aspect of Dirichlet's understanding of analysis. It dealt with the properties of Legendre polynomials, which had also been studied by Poisson using his general method of inserting powers of p into a series and then letting p tend to unity; but Dirichlet pointed out for the first time that Abel's

[25] Dirichlet *1837b*, addition. This addition seems to have been written after the second paper *1837a* on Fourier series, and perhaps under the inspiration of the reworking of the problem. The fact that $F(x)$ *is* continuous was assumed by Dirichlet; it is not difficult to prove via the Bolzano formulation of continuity.

limit theorem showed the fallacy of the method as a proof of convergence of the series involved.[26] So he understood the point of Abel's result; and he must have seen the difficulties of Abel's proof, too, for late in his life he showed to his friend Joseph Liouville (1809–1882) his own proof of the limit theorem when Liouville expressed difficulty in following Abel's reasoning. After Dirichlet's death, Liouville published the proof as a tribute to its creator.

Predictably, it is based on a rearrangement of terms: in fact, the rearrangement implicit in "Abel's partial summation formula"

$$u_0 v_0 + \cdots + u_n v_n = (u_0 - u_1)s_0 + \cdots + (u_{n-1} - u_n)s_{n-1} + u_n s_n, \tag{5.80}$$

where

$$s_r = \sum_{j=0}^{r-1} v_j. \tag{5.81}$$

Liouville had been worried by the part of Abel's proof which showed that the convergence to the sum function took place at the limiting value δ of α. So Dirichlet proved:

THEOREM 5.4
If

$$f(\alpha) = v_0 + v_1 \alpha + v_2 \alpha^2 + \cdots \tag{5.82}$$

is convergent when $0 \leq \alpha < 1$ (where unity serves for Abel's δ), and if

$$A = v_0 + v_1 + v_2 + \cdots, \tag{5.83}$$

then

$$A = f(1). \tag{5.84}$$

From (5.83),

$$s_n = \sum_{r=0}^{n-1} v_r < K \tag{5.85}$$

for some finite constant K. Hence, in (5.82),

$$f(\alpha) = s_0 + \sum_{r=1}^{\infty} (s_r - s_{r-1})\alpha^r \tag{5.86}$$

$$= (1 - \alpha) \sum_{r=0}^{\infty} s_r \alpha^r \tag{5.87}$$

[26] Dirichlet *1837b*, pp. 35–37; *Works*, vol. 1, pp. 285–287. The convergence in question was of the expansion of the generating function, and the polynomials arose from Poisson's own researches in heat diffusion (see esp. Poisson *1835a*, ch. 8).

by an Abelian rearrangement of terms. This is valid because, as Dirichlet pointed out, only the extra term $-s_n \alpha^{n+1}$ is needed to change the nth partial sum of (5.86) into the nth partial sum of (5.87); and this term vanishes with the increase in n, since $\alpha < 1$ and (5.85) shows that s_n is bounded above by K. We now write (5.87) in the "partial sum and remainder" form

$$f(\alpha) = (1-\alpha)\sum_{r=0}^{n-1} s_r \alpha^r + (1-\alpha)\sum_{r=n}^{\infty} s_r \alpha^r \tag{5.88}$$

$$= T_1(\alpha) + T_2(\alpha), \quad \text{say,} \tag{5.89}$$

and, as with the Fourier series, consider the components separately. From (5.85),

$$T_1(\alpha) \leq (1-\alpha)\sum_{r=0}^{n-1} K\cdot 1^r$$

$$= (1-\alpha)nK, \tag{5.90}$$

which vanishes as $\alpha \to 1$. In addition, by a mean value argument,

$$T_2(\alpha) = (1-\alpha)P_n \sum_{r=n}^{\infty} \alpha^r, \tag{5.91}$$

where P_n is within the maximum and minimum of the s_r for $r \geq n$. Therefore

$$T_2(\alpha) = P_n \alpha^n, \tag{5.92}$$

on summation of the geometrical progression in (5.91). We now let $n \to \infty$ and $\alpha \to 1$. From (5.90),

$$\lim_{n\to\infty}\lim_{\alpha\to 1}[T_1(\alpha)] = 0. \tag{5.93}$$

From (5.83),

$$s_n \to A \quad \text{as} \quad n \to \infty \quad \text{when} \quad \alpha = 1, \tag{5.94}$$

and (5.92) shows that $P_n \to A$ also. Therefore in (5.92),

$$\lim_{n\to\infty}\lim_{\alpha\to 1}[T_2(\alpha)] = A. \tag{5.95}$$

Putting (5.93) and (5.95) into (5.89), we see that

$$f(\alpha) \to 0 + A \quad \text{as} \quad \alpha \to 1, \tag{5.96}$$

which proves the theorem as required.[27]

[27] Dirichlet *1862a*. The correspondence between Dirichlet and Liouville is published in Tannery *1908–09a*.

This completes Dirichlet's considerable contribution to analysis: a demonstration of what the new analysis could do if handled with suitable care. Yet there were still problems; Cauchy's theorem 4.4 against Fourier series of discontinuous functions, for example. Dirichlet never referred to this theorem, although his own work had surely refuted it, but he did conjecture that *any* continuous function could be represented by a convergent Fourier series.[28] Cauchy himself stuck to his position even after the publication of Dirichlet's 1829 paper: when he was in Turin in 1833 he published a set of lectures summarizing his Paris teaching, and theorem 4.4 and its proof appeared in it word for word from the *Cours d'analyse*.[29] The crisis over this theorem drifted on into the 1840s: nobody seems to have had any ideas on how to solve it. Certainly the theorem suffered more than the series, for while the study and application of Fourier series remained an important branch of research, the theorem slipped out of succeeding treatises on analysis. Perhaps even the indirect way in which Cauchy had made his point caused some of his successors, ignorant of the circumstances which lay behind it in 1821, to miss seeing the difficulty altogether. At all events, the silence was to be broken in a most remarkable way.

[28] According to a verbal remark by Weierstrass, reported in du Bois Reymond *1876a*, pp. vii–viii. Dirichlet's lectures of the early 1850s on the definite integral were published in 1904. They do not contain any new results of importance to the foundations of integration, but concentrate on the formulation and basic properties of Cauchy's integral, beginning with considerations of necessary and sufficient conditions for the continuity of a function (see Dirichlet *1904a*, esp. arts. 1–27, 39–44, 51–53, 105–106). The same is true of the lectures published by G. F. Meyer in 1871, which were based on Dirichlet's teaching of the late 1850s (see Meyer *1871a*, esp. arts. 1–24, 134–140; and arts. 81–97 for Dirichlet's proof of the convergence of Fourier series).

[29] Cauchy *1833a*, p. 46; *Works*, ser. 2, vol. 10, pp. 55–56. Perhaps this was Cauchy's reply to accurate criticism of his own convergence proof by a protégé of Fourier.

THE AGE OF RIGOR

6

One of Dirichlet's pupils was Phillip Seidel (1821–1896). Seidel's name does not ring through history with the distinction of most of the founders of analysis, for his contributions were not as substantial as theirs; but in 1848 he published a short paper on series which possess a discontinuous sum-function.[1]

"One finds in Cauchy's *Cours d'analyse algébrique*," this paper began, ". . . a theorem which says that the sum of a convergent series, whose individual members are functions of a quantity x and certainly continuous in the neighborhood of a definite value of x, is likewise always a continuous function of the same quantity in this neighborhood. . . .

"Nevertheless the theorem stands in contradiction to what Dirichlet showed, that, for example, Fourier series also always converge if one forces them to develop discontinuous functions;—indeed, the discontinuity will frequently be embedded in the form of these series whose individual members are still continuous functions. . . ."[2]

So Seidel in 1848 was echoing Abel's footnote of 1826 together with the results found by Dirichlet in 1829, and now he was to make his own contribution to the discussion with the following:

THEOREM 6.1

"If one has a convergent series which develops a discontinuous function of a quantity x, whose individual members are continuous functions, then one must be able to assign a value to x in the immediate vicinity of the point where the function jumps for which the series converges *arbitrarily slowly*."[3]

Seidel claimed priority for his new theorem, whose essential difference from Cauchy's lies in the phrase "arbitrarily slowly." From the basic convergence situation

$$s(x) = s_n(x) + r_n(x) \tag{6.1}$$

a noninfinitesimal, but small increment δ on x gives

$$s(x+\delta) - s(x) = [s_n(x+\delta) - s_n(x)] + [r_n(x+\delta) - r_n(x)], \tag{6.2}$$

and the proof of the continuity or discontinuity of $s(x)$ hinges on this equation. Firstly, $s_n(x)$ will be continuous for any finite n, and so

$$|s_n(x+\delta) - s_n(x)| < \varepsilon_1, \tag{6.3}$$

where ε_1 is chosen large enough for the (already selected) increment δ.

[1] Seidel *1848a*.
[2] Seidel *1848a*, pp. 381 and 382.
[3] Seidel *1848a*, p. 383.

In addition, the series is convergent for all relevant values of x, and therefore

$$|r_n(x)| < \varepsilon_2 \tag{6.4}$$

and

$$|r_n(x+\delta)| < \varepsilon_3 \tag{6.5}$$

for appropriately large values of n. Now Seidel advanced beyond Cauchy's reasoning. If $s(x)$ is continuous, then (6.2) shows the need for the smallness not only of $|s_n(x+\delta) - s_n(x)|$ in (6.3), *but also of*

$$|r_n(x+\delta) - r_n(x)|; \tag{6.6}$$

and this has to be achieved by choosing an ε_4 smaller than ε_2 and ε_3 so that (6.4) and (6.5) can apply simultaneously, and considering the smallest integer n_0 for which

$$|r_n(x+\delta)| \leq \varepsilon_4 \quad \text{if} \quad n \geq n_0. \tag{6.7}$$

We decrease δ from an initial value η, and watch the effect on n_0. If it increases to a maximal finite value N as δ passes to 0 then (6.7) will be true for all values of δ between 0 and η when $n > N$, and so (6.4) and (6.5) will be true also. We now make sure that ε_1 in (6.3) is large enough (but still basically "small") for the satisfaction of that equation when $\delta = \eta$ and $n \geq N$ so that (6.3) will apply as well. Hence, using (6.3)–(6.5) in (6.2) we obtain

$$|s(x+\delta) - s(x)| \leq \varepsilon_1 + \varepsilon_2 + \varepsilon_3, \qquad 0 \leq \delta \leq \eta, \tag{6.8}$$

which proves the continuity of $s(x)$ as Cauchy had required. On the other hand, if n_0 increases to infinity as δ approaches zero, then $|r_n(x)|$ must be decreasing "arbitrarily slowly," as Seidel put it: (6.3), and therefore (6.8), will *not* apply and so $s(x)$ will be discontinuous around x.[4]

But we have not finished with new discoveries, for Seidel's paper was slightly preceded by an important but rambling miscellany on analysis in general and Fourier series in particular by another young man: George Stokes (1819–1903).[5] The very beginning of this paper was magnificent, making explicit some of the distinctions which have been discussed earlier but which were still far from generally understood at the time. The difference between Euler's and Cauchy's senses of the term "continuity" was

[4] Seidel *1848a*, pp. 384–388. We have changed Seidel's notation and taken care—where he did not—to use the absolute values of expressions.
[5] Stokes *1847a*.

mentioned, and Stokes made clear his own use of Cauchy's sense. He also followed Cauchy's use of "convergent" and "divergent," to include in the latter category series such as

$$1-1+1-1+\cdots. \tag{6.9}$$

He was aware that this series was the limiting case of the convergent

$$1-g+g^2-g^3+\cdots, \qquad |g|<1, \tag{6.10}$$

but that the two series had to be carefully distinguished. He also noted the subdistinction of convergent series into "essentially" and "accidentally" convergent types, which correspond to the modern terms "absolute" and "conditional," and extended it to infinite integrals and integrals of functions with infinities.[6] In view of Cauchy's theorem 3.5 on integrating a convergent series term-by-term, he surpassed Cauchy in seeing the uselessness of interchanging sum and integral to prove the convergence of the Fourier sine series

$$\frac{2}{a}\sum_{r=1}^{\infty}\int_0^a f(t)\sin\frac{r\pi t}{a}\,dt\,\sin\frac{r\pi x}{a},^{7} \tag{6.11}$$

and borrowed Poisson's idea of treating (6.11) as the limiting value of

$$\frac{2}{a}\sum_{r=1}^{\infty}g^r\int_0^a f(t)\sin\frac{r\pi t}{a}\,dt\,\sin\frac{r\pi x}{a} \tag{6.12}$$

as $g\to 1$; but he surpassed Poisson also in using it only for the purpose of finding the sum of the series, given its convergence.[8] Convergence itself was demonstrated by what he himself described as a "rather circuitous" proof based on integrating the general coefficient in (6.11) by parts, and effectively carrying out an analytical equivalent of Dirichlet's geometrical reasoning.[9] One begins to wonder at the purpose of this analysis in view

[6] Stokes *1847a*, art. 1.

[7] Stokes *1847a*, art. 2. Stokes quoted theorem 3.5 from its appearance in the treatise on the calculus by Cauchy's apprentice the abbot F. N. M. Moigno (see Moigno *Calculus*, vol. 2 (1844), pp. 70–71), who devoted his career to the development of teaching of pure and applied mathematics according to Cauchy's principles.

[8] Stokes *1847a*, arts. 3–5.

[9] Stokes *1847a*, arts. 6 and 7. Stokes would appear not to have made full use of his method, for his expression (11) can be used to seek conditions of convergence for a function with an infinite number of discontinuities. The expression (11) shows that we may write our (6.11) as

$$\frac{2}{\pi}\sum_{J=0}^{m}\left\{[f(d_J+0)-f(d_J-0)]\sum_{r=1}^{\infty}\left[\frac{1}{r}\cos\frac{r\pi\,d_J}{a}\sin\frac{r\pi x}{a}+0\left(\frac{1}{r^2}\right)\right]\right\} \tag{1}$$

where $d_0=0$, $d_m=a$, and $d_1,\ldots,d_{m-1}$ are the abscissas of the points of discontinuity of $f(x)$. If m is infinite, the j-summation in (1) becomes infinite and the convergence question reduces to the question of the convergence of (1).

There is a slight reminiscence in this procedure of Dirksen's proof-method for the convergence of Fourier series (see n. 22, p. 105).

of Dirichlet's achievements, when a footnote explains the situation: Stokes had not been aware of Dirichlet's work when he had carried it out.[10] The new knowledge seemed to do him little good, however, for he launched himself into a long and obscure discussion of Fourier series from every point of view except convergence.[11] But in the next section he produced a remarkable new result in the convergence of infinite series in general.

Stokes had been particularly concerned with the value of the Fourier series at a point of discontinuity of the function. So his problem may be put as follows: let

$$\sum_{r=1}^{\infty} v_r(h) = V(h) \quad \text{when} \quad 0 < h < \varepsilon, \tag{6.13}$$

and

$$\sum_{r=1}^{\infty} v_r(0) = U. \tag{6.14}$$

Under what conditions does

$$V(0) = U? \tag{6.15}$$

The problem is that to which Abel had offered his limit theorem for the particular case of a power series. Stokes saw at once that it was concerned with the validity of a double limit, for, if we write $s_n(h)$ for the nth partial sum of $V(h)$,

$$\lim_{h \to 0} \lim_{n \to \infty} [s_n(h)] = V(0) \tag{6.16}$$

and

$$\lim_{n \to \infty} \lim_{h \to 0} [s_n(h)] = U, \tag{6.17}$$

and so (6.15) expresses the validity of interchanging the order of limit-taking.

[10] Stokes *1847a*, footnote to art. 7. Stokes's ignorance of Dirichlet's work and awareness of Poisson's is characteristic of British mathematicians, who lagged well behind Continental developments in analysis all through the century. Despite the attention that was given to Poisson's proof from time to time (see n. 5, p. 90), it became rapidly disowned on the Continent in favor of Dirichlet's; but even as late as 1893, G. Gibson reported that it was still normally used in English teaching of the convergence problem (see Gibson *1893a*, pp. 147–148). A good example of the lack of understanding of analytical developments at the time of their creation can be gathered from the rambling report made to the British Association in 1834 by G. Peacock. (See Peacock *1834a*. For commentary on British analysis during this period, see Hardy *1949a*, pp. 18–20.) Apparently ideas were no more in ferment in Belgium (see Bockstaele *1966a*).

[11] Stokes *1847a*, arts. 9–27. In arts. 28–37 he similarly treated Fourier integrals.

The problem seems similar to Seidel's; and when we realize that Stokes was assuming that each $v_r(h)$ is a continuous function we see that the problem is of exactly the same type, for the extension of (6.13) to the case $h = 0$ is a question of the continuity of $V(h)$ as $h = 0$ is approached from the right-hand side. Further, although he did not mention Cauchy's theorem 4.4 of the *Cours d'analyse* and indeed may not have been aware of it,[12] his solution to the problem was strangely similar to and different from Seidel's. For he proved

THEOREM 6.2

$V(0) = U$ (and so $V(h)$ is a continuous function of h at $h = 0$) unless the convergence of (6.13) becomes "infinitely slow when h vanishes."[13]

So Stokes, too, had seen that the convergence of the series had to be of a special kind for the continuity of the sum function. His terminology is strikingly similar to Seidel's and, as we have said, the problem is the same; but in fact his solution is fundamentally different from Seidel's, and this circumstance is sufficient reason in itself for dismissing any suggestion that Seidel knew of Stokes's paper.[14] As a contrast to Seidel's "arbitrarily slow" convergence we quote from Stokes:

DEFINITION 6.1

"The convergency of the series is here said to become infinitely slow when, if n be the number of terms which must be taken in order to render the sum of the neglected terms numerically less than a given quantity e which may be as small as we please, n increases beyond all limit as h decreases beyond all limit."

[12] Cauchy's theorem is not mentioned in the volumes of Moigno's *Calculus* of which Stokes was aware (see n. 7, p. 114), and he may well not have gone back to Cauchy's own books.

[13] Stokes *1847a*, art. 39. The theorem shows another affinity with the thought of Dirksen. A few months after his effort on the convergence of Fourier series in 1829 (see n. 22, p. 105), Dirksen reviewed the recently published German translation by C. Huzler of Cauchy's *Cours d'analyse*. Remarking on the falsity of theorem 4.4 following Abel's footnote of 1826, he suggested that a condition for validity could be provided by—nothing other than the equality of the double limits (6.16) and (6.17). However, he took the idea no further (see Dirksen *1829c*, cols. 217–218).

[14] Stokes read his paper to the Cambridge Philosophical Society in December, 1847, and it was published in the Society's *Transactions* in 1849. Seidel's paper was published in the *Abhandlungen der Bayerische Akademie der Wissenschaften* at Munich during 1848. Therefore Seidel's paper was written before Stokes's was published; and, indeed, the tone of Seidel's work gives the impression of genuinely original research. There is one odd coincidence, in that both include some consideration for the Fourier integral $\int_0^\infty t^{-1} \sin xt\, dt$, but Stokes treated it *as* a Fourier integral, while Seidel quoted it as an example of a discontinuous function satisfying Dirichlet's conditions (see Stokes *1847a*, art. 41; Seidel *1848a*, p. 391).

A consequence of this definition is that "if the convergency do not become infinitely slow, it will be possible to find a number n_0 so great that for the value of h we begin with and for all inferior values greater than zero the sum of the neglected terms shall be numerically less than e."[15]

Symbolically we may express Stokes's "not infinitely slow convergence" as follows: we take the series

$$s(x) = \sum_{r=1}^{\infty} u_r(x) \tag{6.18}$$

$$= s_n(x) + r_n(x) \tag{6.19}$$

in the usual way. Then, writing ε for e, his condition is that:

Given the small quantities ε and h, there exists a number n_0 such that

$$|r_n(x)| < \varepsilon \quad \text{if} \quad n = n_0 \quad \text{and} \quad 0 \leq x \leq h. \tag{6.20}$$

However, Seidel's imposition on $|r_n(x)|$ in (6.4)–(6.8) is different. In fact, he reduces the increment δ of x rather than ε of $f(x)$, but the procedure is equivalent in that ε can (but need not) be reduced at the same time, and the effect on $|r_n(x)|$ can be written as follows:

Given the small quantities ε and h, there exists a number n_0 such that

$$|r_n(x)| < \varepsilon \quad \text{if} \quad n \geq n_0 \quad \text{and} \quad 0 \leq x \leq h. \tag{6.21}$$

The difference seems small; but it is crucial. Seidel's n_0 is dependent on ε for $0 \leq x \leq h$, and then he requires that $|r_n(x)| < \varepsilon$ for *all* larger values of n. However, Stokes's condition applies *only at* $n = n_0$: $|r_n(x)|$ is *not* necessarily $< \varepsilon$ for values of n larger than n_0. The point is more important than it may at first appear, and to appreciate it we must leave both Stokes and Seidel for the man who actually anticipated them in this type of discovery and overtook them in understanding its significance—Karl Weierstrass (1815–1897).

The earliest relevant manuscript of Weierstrass dates from 1841, but it deals not with Cauchy's theorem but with the convergence of power series of the form $\sum_{r=-\infty}^{+\infty} a_r x^r$.[16] For the fault with Cauchy's theorem is not a minor slip of reasoning, but a fundamental lack of understanding of a vital aspect of analysis of which the theorem was only a manifestation, and which Weierstrass began to reveal only when he began to lecture at

[15] Stokes *1847a*, art. 39. We have written n_0 for his n_1.

[16] Weierstrass *1841a*; see esp. pp. 68–69. The manuscript was first published in his Collected Works.

Berlin in the 1850s after years of obscurity as a schoomaster.[17] That aspect is *the convergence of a series of functions* as opposed to a series of constant terms. Cauchy knew how to handle the convergence of terms, but he never realized that when these terms were functions of a variable x then the clash between his theorem and Fourier series was to be resolved by introducing distinctions between *different modes of convergence relative to the variable* x. Thus he missed not only the "not infinitely/arbitrarily slow" modes of convergence that his theorem needed, but equally the "infinitely/arbitrarily slow" modes that Fourier series of discontinuous functions needed. Stokes and Seidel surpassed him in seeing that such distinctions were necessary; and Weierstrass surpassed them both in the use of them and especially in their basic formulation. Following him, we speak today of modes of "uniform" and "nonuniform" convergence, and the difference of terminology reflects a more profound appreciation of the problem. Let us contrast the two approaches:

DEFINITION 6.2 (Stokes-Seidel)
The convergence is $^{\textit{not infinitely}}_{\textit{infinitely}}$ *slow over* $[a, b]$ if, given that $|r_n(x)| < \varepsilon$, the index n $^{\text{tends}}_{\text{does not tend}}$ to infinity as ε tends to zero when $a \leq x \leq b$.

A difficult idea to handle, especially arithmetically. Contrast it with

DEFINITION 6.3 (Weierstrass)
The convergence is $^{\textit{uniform}}_{\textit{nonuniform}}$ *over* $[a,b]$ if, for a given ε, there $^{\text{is}}_{\text{is not}}$ an n for which $|r_n(x)| < \varepsilon$ when $a \leq x \leq b$.

Thus Weierstrass had done no more than insert the variable x back into Bolzano's basic formulation of convergence after Cauchy had omitted it, and in handling modes of convergence *preserved the limit-avoiding and arithmetical character of the reasoning* in a way that Stokes and Seidel had muddled by introducing limit-*achieving* elements with the behavior of n as ε achieved zero. But there is still more: we recall that Stokes and Seidel had suggested different modes within their common approach of "(not) infinitely/arbitrarily slow" convergence. These two modes are far from exhaustive for series of functions: following G. H. Hardy (1877–1947) we may classify the principal ones—in Weierstrassian terms—as follows:

[17] For information on his life, and correspondence, see the articles in *Acta Mathematica*, vol. 39 (1923); Mittag-Leffler *1902a* and *1912a*; and Biermann *1966a*. The latter also contains a valuable list of writings on Weierstrass up to 1966 (see pp. 218–220).

DEFINITION 6.4

Mode A1: $\substack{\textit{Uniform}\\ \textit{Nonuniform}}$ *convergence over an interval* $[a, b]$. Given any positive order of smallness ε, there $\substack{\text{exists}\\ \text{does not exist}}$ an integer n_0 such that

$$|r_n(x)| < \varepsilon \quad \text{if} \quad n \geq n_0, \qquad a \leq x \leq b. \tag{6.22}$$

DEFINITION 6.5

Mode A2: $\substack{\textit{Uniform}\\ \textit{Nonuniform}}$ *convergence in a neighborhood of the value* x_1 *of* x. There $\substack{\text{exists}\\ \text{does not exist}}$ a positive δ for which the series is uniformly convergent over $[x_1 - \delta, x_1 + \delta]$.

DEFINITION 6.6

Mode A3: $\substack{\textit{Uniform}\\ \textit{Nonuniform}}$ *convergence at the point* x_1. Given any positive order of smallness ε, there $\substack{\text{exists}\\ \text{does not exist}}$ a neighborhood $[x_1 - \delta, x_1 + \delta]$ and an integer n_0 for which (6.22) applies.[18]

Weierstrass was aware of modes A1 and A2: Seidel's brand, as we can see, was also mode A2. But Stokes's version belongs to the second group:

DEFINITION 6.7

Mode B1: Quasi-$\left\{\substack{\textit{uniform}\\ \textit{nonuniform}}\right.$ *convergence over an interval* $[a, b]$. Given any positive order of smallness ε, there $\substack{\text{exists}\\ \text{does not exist}}$ an infinity of integers n_0 such that

$$|r_n(x)| < \varepsilon \quad \text{when} \quad n = n_0, \qquad a \leq x \leq b. \tag{6.23}$$

DEFINITION 6.8

Mode B2: Quasi-$\left\{\substack{\textit{uniform}\\ \textit{nonuniform}}\right.$ *convergence in a neighborhood of the value* x_1 *of* x. There $\substack{\text{exists}\\ \text{does not exist}}$ a positive δ for which the series is quasi-uniformly convergent over $[x_1 - \delta, x_1 + \delta]$.

DEFINITION 6.9

Mode B3: Quasi-$\left\{\substack{\textit{uniform}\\ \textit{nonuniform}}\right.$ *convergence at the point* x_1. Given any positive order of smallness ε, there $\substack{\text{exists}\\ \text{does not exist}}$ a neighborhood $[x_1 - \delta, x_1 + \delta]$ and an infinity of integers n_0 for which (6.23) applies.[19]

Stokes was invoking mode B2, with only one of the values n_0 of n stipulated for which $|r_n(x)|$ is suitably small. An important feature of post-Weierstrassian analysis was the development of these modes and the investigation of the logical relations between them. Clearly, each mode of

[18] See Hardy *1918a*, art. 3.

[19] See Hardy *1918a*, art. 5. There are still more sophisiticated modes of uniform and nonuniform convergence, mainly of a set-theoretic character (see Hardy *1918a*, art. 6).

uniform convergence implies the corresponding mode of quasi-uniform convergence; and within uniform convergence Seidel's mode A2 implies both A1 and A3.[20] Cauchy's theorem 4.4 from the *Cours d'analyse* is therefore true for all modes of uniform and quasi-uniform convergence and false otherwise. Functions satisfying Dirichlet's conditions possess uniformly or nonuniformly convergent Fourier series according as they are also continuous or not.[21]

This is the kind of analysis with which we are now familiar: the analysis of "the age of rigor." But the age did not dawn before Weierstrass, for these levels of technique and subtlety of reasoning were introduced only in his analysis lectures at Berlin. Uniform and nonuniform convergence; noninfinitesimal analysis to avoid the difficulties of his infinitesimalist predecessors; the "(ε, δ)" formulation of Bolzano's arithmetical approach to analysis: all these ideas were urged by Weierstrass on his students, who then began to use and develop them in their own research and teaching. Weierstrass himself did not publish his analysis lectures and discouraged even the taking of notes, but he seems to have helped to create the legend of Cauchian rigor by carrying out Cauchy's aspiration that "as for methods, I have sought to give them all the rigor that one demands in geometry so as never to resort to reasoning taken from the generality of algebra."[22]

Weierstrass undoubtedly saw himself as Cauchy's heir in analysis and so helped to create the belief that Cauchy's achievements included ideas that were actually his own. However, he had begun his revision of analysis

[20] For proofs of these relations and also those holding between modes B1, B2 and B3, see Hardy *1918a*, arts. 4 and 6. For commentary on Stokes and Seidel, see Reiff *1889a*, art. 17. In his paper, Stokes tried to prove the converse of his theorem 6.2 and hence establish mode B2 as the necessary and sufficient condition for the continuity of the sum-function (*1847a*, art. 39); but he actually used mode B3 (quasi-uniform convergence at a point), which does in fact provide the condition for which he was looking (see Hardy *1918a*, art. 8).

[21] Uniform and nonuniform convergence is badly presented in teaching literature. The problem itself is usually not mentioned at all, and instead some property called "uniform convergence" (of mode A1) is invoked from time to time to prove various theorems, with nonuniform convergence left out of the picture altogether. Hence the fundamental importance of the problem for analysis, which Weierstrass emphasized in his own teaching (according to Stolz *1881a*, p. 256), is lost.

[22] Cauchy *1821a*, introduction, p. ii. (For some general remarks on Cauchy's achievements in analysis, see Carruccio *1957a*.)

Some idea of Weierstrass's teaching of analysis may be gathered from the notes published in 1880 by S. Pincherle "following Weierstrass's principles" in his Berlin lectures of the late 1870s (Pincherle *1880a*). Lecture courses on other branches of mathematics were included in the edition of Weierstrass's Collected Works (see Weierstrass *Works*, vols. 4–7).

before he read any Cauchy;[23] and when he did, he discovered another flaw in Cauchy's analysis involving uniform convergence, for he found that Cauchy's theorem 3.5 on integrating a convergent series term-by-term actually assumed the uniform convergence (mode A1) of the series over $[\alpha_0, X]$. This assumption may be easily detected in Cauchy's own proof; but, in contrast to his theorem 4.4 of the *Cours d'analyse*, counterexamples were not at all obvious and so the theorem was not under attack. We give Cauchy's proof together with the required modification:

THEOREM 6.3
If $\Sigma_{r=0}^{\infty} u_r(x)$ converges uniformly to $s(x)$ over $[x_0, X]$, then we may integrate the series term-by-term. Symbolically, if

$$\sum_{r=0}^{\infty} u_r(x) = s(x), \qquad x_0 \leq x \leq X, \tag{6.24}$$

then

$$\sum_{r=0}^{\infty} \int_{x_0}^{X} u_r(x)\, dx = \int_{x_0}^{X} s(x)\, dx = \int_{x_0}^{X} \sum_{r=0}^{\infty} u_r(x)\, dx. \tag{6.25}$$

From (6.24),

$$s(x) = \sum_{r=0}^{n-1} u_r(x) + r_n(x) \tag{6.26}$$

and so

$$\int_{x_0}^{X} s(x)\, dx = \sum_{r=0}^{n-1} \int_{x_0}^{X} u_r(x)\, dx + \int_{x_0}^{X} r_n(x)\, dx. \tag{6.27}$$

From the mean value theorem (theorem 3.3),

$$\int_{x_0}^{X} r_n(x)\, dx = (X - x_0) r_n(x_1), \qquad x_0 < x_1 < X. \tag{6.28}$$

Now whatever be the point x_1, the series will be convergent there, by assumption: therefore $r_n(x_1)$ will be small, and so $\int_{x_0}^{X} r_n(x)\, dx$ will be small also. But Cauchy failed to realize that, as an integrand, $r_n(x)$ would have to be within that order of smallness *throughout the interval* for $\int_{x_0}^{X} r_n(x)\, dx$ also to be small in (6.28). This is what Weierstrass saw; he noticed that it amounted to an assumption of uniform convergence (mode A1) in the

[23] Weierstrass was using uniform convergence in 1841 (see n. 16, p. 117), but he did not begin to read Cauchy until 1842 (according to Mittag-Leffler *1923a*, pp. 35–36).

original series (6.24). Having made this assumption, the proof of the theorem follows easily from (6.27).[24] From it we may also develop a theorem on the validity of differentiating a series term-by-term in terms of the uniform convergence of the series of derivatives, which would improve upon Cauchy's treatment of the problem in (3.23)–(3.28); and it seems quite possible that Weierstrass also gave the result in his Berlin lectures.

Cauchy lived through the appearance of modes of convergence and returned to his theorem 4.4 in a short paper of 1853, five years after Seidel's explicit reference to and solution of the problem. As usual, Cauchy mentioned only support for his original theorem in the special case of a power series (where it is of course validated by Abel's limit theorem), but then he admitted that

$$\frac{\pi - x}{2} = \sin x + \frac{1}{2}\sin 2x + \frac{1}{3}\sin 3x + \cdots \tag{6.29}$$

—nothing other than a Fourier series, in fact!—was a counterexample with its discontinuity of magnitude π when x equals a multiple of π.[25] So a repair was needed, precisely at the point in his original proof of theorem 4.4 where he had said that $[r_n(x_0 + \alpha) - r_n(x_0)]$ "becomes insensible at the same time" as $r_n(x_0)$. This remark is difficult to interpret against the classification of modes of uniform convergence given here (having been made before its development), since α is an infinitesimally small increment of x. We may describe it as mode A2 over an infinitesimal neighborhood of x_0, but different from mode A3 at a point since α is independent of the "insensibility of smallness" to which $[r_n(x_0 + \alpha) - r_n(x_0)]$ tends along with α. At all events, Cauchy did not undertake modifications to his old proof, but stated a new theorem which embodied the extension of the necessary and sufficient conditions for convergence, and a revised theorem 4.4 to cover uniform convergence of mode A1:

[24] Cauchy's version is in Cauchy *1823a*, lecture 40. E. Heine reports that Weierstrass first gave the correct version (see Heine *1870a*, p. 353). On Weierstrass's contributions to pure mathematics, see Jourdain *1909a*.

[25] Cauchy *1853a*, p. 455; *Works*, ser. 1, vol. 12, p. 31. Cauchy's supporters, cited only by surname in the paper, were A. A. Briot and J. C. Bouquet, who made important use of Cauchy's analysis in their various books and papers on elliptic functions. Apparently they were not friendly to Cauchy outside research (according to Bertrand *1904a*, p. cxci).

THEOREM 6.4

If the $u_r(x)$ are each continuous over $[x_0, X]$ and if $|s_{n'}(x) - s_n(x)|$ "becomes always infinitely small for infinitely large values of the numbers n and $n' > n$, the series $[\Sigma_{r=0}^{\infty} u_r(x)]$ will be convergent, and the sum $[s(x)]$ of the series will be a continuous function of the variable x between the given limits."[26]

Apart from the handling of infinitesimals, the proof of the theorem is fairly clear: the combined smallness of $r_n(x_0)$ and $r_n(x_0 + \alpha)$ by uniform convergence, and $[s_n(x_0 + \alpha) - s_n(x_0)]$ by continuity of $s_n(x)$, give the continuity of $s(x)$ from

$$s(x_0 + \alpha) - s(x_0) = [s_n(x_0 + \alpha) - s_n(x_0)] + [r_n(x_0 + \alpha) - r_n(x_0)]. \qquad (6.30)$$

But there was much confusion of presentation. The necessity of the condition was proved before the revised theorem, but the sufficiency was dealt with afterwards. Absolute values of expressions were used in both parts of the condition, but not mentioned in his own statement of the theorem nor used in the revision of the result from the *Cours d'analyse*, while uniform convergence itself was tucked away in the word "always" with no reference to the variable at all. The whole business of this theorem and its revision was of great importance for analysis, but from Cauchy's point of view it was an unhappy affair. In 1853, near the end of his life, he seemed outdated: the center of attention in mathematics had moved from Paris to Berlin and Göttingen. In both of these cities moved a young man who was to set the next milestone in analysis—Bernhard Riemann (1826–1866).

In 1854, Riemann submitted three papers to the University of Göttingen for his "inauguration" examination for a university lectureship. Three years previously he had offered his masterpiece on the theory of functions of a complex variable for his doctorate.[27] Gauss had been very impressed with it, as it dealt with a branch of mathematics which had never properly left his notebooks.[28] Now Gauss chose another of his unpublished interests from Riemann's three titles: non-Euclidean geometry. Riemann scored an even greater success with his paper, for Gauss was very pleased. One

[26] Cauchy *1853a*, pp. 456–457; *Works*, ser. 1, vol. 12, p. 33. Later P. du Bois Reymond generalized the theorem to the following: If $\Sigma_{r=0}^{\infty} \mu_r$ is an absolutely convergent series and $u_r(x)$ is a continuous finite function over $[a, b]$ then $\Sigma_{r=0}^{\infty} \mu_r u_r(x)$ is also continuous over $[a, b]$. (See du Bois Reymond *1871a*.)

[27] Riemann *1851a*.

[28] See n. 14 and text, p. 30 on Gauss's unpublished work.

may wonder how Gauss would have reacted to another title, which had been Riemann's original intention for his inauguration: *Über die Darstellbarkeit einer Function durch eine trigonometrische Reihe* (*On the developability of a function by a trigonometric series*).

Riemann had shown a draft of this paper to Dirichlet in 1852, who then acquainted him with his own reminiscences of the problem.[29] On Gauss's death in 1855, Dirichlet took his chair; after Dirichlet's death in 1859, Riemann himself succeeded; yet on his own death in 1866 his manuscript on trigonometric series still remained unpublished, and so his close friend Richard Dedekind (1831–1916) brought it out in 1866, along with the paper on non-Euclidean geometry which had given Gauss so much pleasure.[30] In view of the encouragement given to the inauguration paper, it is surprising that Riemann did not publish it himself; but the paper on trigonometric series was unfinished in some aspects. Nevertheless, he had made such progress on the problem that its publication was a landmark in the history of analysis, for it ushered in the new era of Riemannian research in the subject. The study of this era needs a separate volume,[31] but we can discuss those parts of Riemann's paper which rounded off the Cauchian period; for, like the Dirichlet papers on Fourier series, it included general considerations on the state of analysis at the time. In fact, it began with a superficial historical survey of trigonometric series from the vibrating string problem through Fourier to Dirichlet, where Riemann recorded Dirichlet's motivation to the convergence problem in the rearrangement of conditionally convergent series.[32] Immediately he generalized Dirichlet's point. If we take a_r to represent the rth positive term of the series and $-b_r$ to be the rth negative term, then $\Sigma_{r=1}^{\infty} a_r$ and

[29] See the letter from Riemann to his father quoted in Dedekind *1876a* in Riemann *Works*$_1$, p. 514; *Works*$_2$, p. 546.

[30] Riemann *1866a* and *1866b*. Dedekind also reprinted *1851a* (see n. 27, p. 122) in 1867, and was largely responsible for the first edition of Riemann's Collected Works in 1876, for which the nominal editor was W. Weber (Riemann *Works*$_1$: see Weber's preface). Dedekind and Weber issued a second edition in 1892, containing some additional unpublished fragments whose number was greatly extended in a supplement of 1902 edited by M. Nöther and W. Wirtinger (Riemann *Supplement*).

[31] The best accounts of those developments—while still being incomplete in many ways—are to be found in Schönflies *Sets*, Schönflies and Baire *1909a*, and Medvedev *1965a* (mainly on set theory); Borel *1912a* (set and measure theory, and convergence of series of functions); Pesin *1966a* (measure theory and integration); Grattan-Guinness *1970a* (manuscripts of Cantor on set theory); and Jourdain *1906a* (theory of functions).

[32] Riemann *1866a*, arts. 1–3. We have criticized this account in n. 27, p. 12 and n. 44, p. 19. From Riemann's letter cited in n. 29 above, it seems possible to conclude that his meeting with Dirichlet motivated him to write this historical account in the first place.

$\Sigma_{r=1}^{\infty} b_r$ are both necessarily divergent series if the original series is to be conditionally convergent. Therefore we can rearrange the terms to give *any* sum C of our choosing, for we would merely have to take enough of the a_r (in order) to surpass C, then enough of the $-b_r$ to fall below it, then enough of the rest of the a_r to surpass C again, and so on. Now the terms of the original series, and therefore a_r and b_r also, tend to zero as r tends to infinity: therefore the increments and decrements around C decrease as well, and so the rearranged series converges to that value.[33]

Riemann's main aim was to extend Dirichlet's results on the convergence of the series: to find conditions for convergence involving an infinity of turning values and discontinuities in the function, and the value that the convergent series would take. Significantly, he felt that Dirichlet had exhausted the possibilities of the direct approach based on the nth partial sum, and so he examined the second derivative of the second integral of the series:

$$A + Bx - \frac{1}{2} a_0 x^2 - \sum_{r=1}^{\infty} \left[\frac{1}{r^2} (a_r \cos rx + b_r \sin rx)\right] \tag{6.31}$$

where a_0, a_r, b_r, are the Fourier coefficients of $f(x)$ over $[-\pi, +\pi]$. By this means he advanced beyond Dirichlet—even if not as far as he intended—and realized that the integrals of some of his new functions, which would appear in the Fourier coefficients, were not easy to formulate in terms of the limit of the sum. So he introduced a variant of Cauchy's argument, based on the oscillation of the (finite-valued) function within each sub-interval of the partition, and found a necessary and sufficient condition for the Cauchy-integrability of a function.

Let us partition the interval with the points

$$x_0(= a), x_1, x_2, \ldots, x_{n-1}, x_n(= b), \tag{6.32}$$

and write

$$\delta_r = x_r - x_{r-1}. \tag{6.33}$$

If we assume that $f(x)$ is Cauchy-integrable, then the sum

$$S = \sum_{r=1}^{n} f(x_r + \varepsilon_r \delta_r)\, \delta_r, \qquad 0 \le \varepsilon_r \le 1, \tag{6.34}$$

33 Riemann *1866a*, art. 5. Riemann tried to give the impression that the argument is Dirichlet's, but there is little doubt that it is his own.

converges to a unique limit as the partition becomes finer. We define the oscillation D_r of $f(x)$ over $[x_{r-1}, x_r]$ by taking the difference of

$$M_r = \max f(x), \qquad x_{r-1} \leq x \leq x_r \tag{6.35}$$

and

$$m_r = \min f(x), \qquad x_{r-1} \leq x \leq x_r \tag{6.36}$$

so that

$$D_r = M_r - m_r; \tag{6.37}$$

and then we form

$$P = \sum_{r=1}^{n} D_r \delta_r . \tag{6.38}$$

Let

$$\Delta = \max P, \quad \delta_r < d, \qquad r = 1, 2, \ldots, n \tag{6.39}$$

for a given (small) magnitude d, and

$$s = \Sigma\,(x_r - x_{r-1}), \tag{6.40}$$

where the summation is taken for all the subintervals of the partition (6.32) of $[a, b]$ for which

$$D_r \geq \sigma. \tag{6.41}$$

From (6.39)–(6.41) the collective contributions to P in (6.38) by these subintervals is clearly $\geq \sigma s$. Thus, *a fortiori*, P itself is greater than σs and so, from (6.39),

$$\sigma s \leq P \leq \Delta \tag{6.42}$$

and

$$s \leq \frac{\Delta}{\sigma}. \tag{6.43}$$

Now the assumption that S tends to a unique limit implies that P tends to zero. Therefore, from (6.39), Δ tends to zero also, and so from (6.43), so does s. Thus Riemann had proved

THEOREM 6.5
"In order that the sum S converges, if collectively the δ become infinitely small, besides the finitude of the functions $f(x)$ it is still necessary that the total size of the intervals in which the oscillations are $>\sigma$, whatever σ may be, can be made arbitrarily small through appropriate values of d."[34]

Riemann saw that the converse was also true. From (6.40), the contribution to P from the subintervals with oscillation greater than σ will be less than

$$Ds, \tag{6.44}$$

where D is the oscillation of $f(x)$ over $[a, b]$, while the contribution from the other intervals will be less than

$$\sigma(b-a). \tag{6.45}$$

Thus, from (6.44) and (6.45),

$$P < Ds + \sigma(b-a). \tag{6.46}$$

The quantity σ is small by choice, and when we assume the condition of theorem 6.5, s will be small also. Hence P tends to zero, and therefore S tends to a unique limit, as the fineness of partitioning increases, and we have theorem 6.5 and its converse together:

THEOREM 6.6
A necessary and sufficient condition for the Cauchy-integrability of the finite-valued function over $[a, b]$ is that the total length of the subintervals for which the oscillation is greater than any given quantity σ is arbitrarily small.[35]

Riemann's basic approach is very old: as he said of it himself in his lectures of the 1850s on partial differential equations: "yes, other things being equal, Archimedes himself has this method for his calculation of curved lines."[36] Yet even though the result is retrospective over Cauchy's analysis, his proof-method contains astonishing anticipations of the

[34] Riemann *1866a*, art. 5.
[35] Riemann *1866a*, art. 5. For infinities in the integrand Riemann used Cauchy's "singular integral" technique, as usual.
[36] Riemann *Lectures*, p. 9. The lectures were first published in 1869.

reasoning that his work on Fourier series was to inaugurate.[37] The condition itself states that the set of points of discontinuity of the function is to be of measure zero, and the set of points in (6.40) is assumed to be measurable for all σ. No wonder the paper caused such a stir—even today his condition is usually given in the equivalent and more homely form due to Gaston Darboux (1842–1917) in 1875:

DEFINITION 6.10
From (6.35) and (6.36) the *upper* and *lower sums* of $f(x)$ over $[a, b]$, U and L, are given by

$$U = \sum_{r=1}^{n} M_r \delta_r \tag{6.47}$$

and

$$L = \sum_{r=1}^{n} m_r \delta_r . \tag{6.48}$$

[37] We may mention here the principal efforts to extend Dirichlet's conditions before the publication of Riemann's paper. In 1839, Bessel took the nth partial sum for all n expressible in the form $(k^3 - 1)/2$, where k is an integer, developed the Dirichlet integral to introduce powers of $1/k$, and then let $k \to \infty$; but the argument seems to add nothing to Dirichlet's work (see Bessel *1839a*). Bessel's motivation to Fourier series is of historical interest. Just as Fourier developed the basic theory of Bessel functions years before Bessel (see Grattan-Guinness *1969a*, pp. 245–247), so Bessel came to Fourier series (through problems of theoretical astronomy) during the 1810s independently of Fourier: indeed he stated the series in a paper of 1816 with reference neither to its derivation nor to any of the mathematical problems involved (see Bessel *1816a*, pp. 49–50; *Papers*, vol. 1, p. 18). He may well have learned of the series from Gauss, who seems to have known of it himself in 1809 (see the examples in the manuscript Gauss *1809a*), in the course of their extensive correspondence. (The surviving letters are published in Gauss *1880a*. A letter from Gauss in September 1805 mentions trigonometric expansions: see pp. 10–11.)

In 1850, O. Bonnet offered an alternative version of Dirichlet's proof based on theorems of the mean value type, for which he was later to become well-known (Bonnet *1850a*, pp. 3–25). But the most significant contribution was a paper of 1864 by R. Lipschitz, where he tried to fulfill Dirichlet's promise of 1829 of finding the "some other quite remarkable properties" of Fourier series (see n. 22 and text, p. 105). One possibility for extension was to an infinity of discontinuities, but Lipschitz could not make any progress with it (Lipschitz *1864a*, pp. 298–299). However, to cover the case of an infinity of oscillations of decreasing magnitude around a point he offered the "Lipschitz condition"

$$|f(x+\delta) - f(x)| < B\,\delta^{\alpha}, \qquad \alpha > 0, \tag{1}$$

with which he replaced Dirichlet's finite number of turning values and so extended Dirichlet's conditions for convergence of the series. His remarks on the generality of his condition imply some aspects of post-Riemannian analysis (Lipschitz *1864a*, pp. 299–308; for commentary see Sachse *1880a*, pp. 244–249). Like Dirichlet, Lipschitz promised another paper on the subject: as with Dirichlet, it never appeared.

THEOREM 6.7
A necessary and sufficient condition for the Cauchy-integrability of $f(x)$ over $[a, b]$ is that for a given small quantity ε,

$$|U-L| < \varepsilon \tag{6.49}$$

for a suitably fine partition of $[a, b]$.[38]

Soon Weierstrass's pupils were all working on problems in analysis inspired by Riemann: infinitely oscillatory and/or discontinuous functions; continuous nondifferentiable functions; modes of uniform and nonuniform convergence; point discontinuities of Fourier series: and so on. This was the 1870s, the time of Hankel's contemporaries: the age of Bolzano's "pure analysis" had arrived with a vengeance.

[38] Darboux *1875a*, sect. 2. See also the contemporary paper Smith *1875a*.

APPENDIX
THE SEARCH FOR CONVERGENCE TESTS

". . . before effecting the summation of any series," wrote Cauchy in the introduction to the *Cours d'analyse*, "I had to examine in which cases the series can be summed, or, in other words, what are the conditions of their convergence; and on this topic I have established general rules which seem to me to merit some attention."[1]

Having learned to interpret the summation of a series solely in terms of the behavior of its nth partial sum and the necessary and sufficient condition for convergence, Cauchy realized that that condition would in practice be difficult to implement, and so he created a new problem which Bolzano did not tackle but which formed an important application of his "pure analysis": the discovery of sufficient conditions, or "tests," for a series to be convergent. The question is quite independent of, and indeed prior to, the summation of the series, and an explanation of its logical structure will serve as a valuable background for the understanding of the results obtained.

DEFINITION A.1

Suppose we have two mathematical statements P and Q in the logical relation

$$P \rightarrow Q. \tag{A.1}$$

Then we say that P is a *sufficient condition* for the occurrence of Q, while Q is a *necessary condition* for the occurrence of P.

$P \rightarrow Q$ is best phrased as "P implies Q," or "if P, then Q," or "Q only if P"; but unfortunately it often is described as "Q if P," which is very easily confused with $Q \rightarrow P$. The proof that in fact $P \rightarrow Q$ is the task of mathematics: the logical point is the *interpretation* of $P \rightarrow Q$, which depends on whether sufficient conditions for Q or necessary conditions for P are being sought. Possibly we are seeking necessary as well as sufficient conditions, which in the case of Q would correspond to the logical pattern

$$P \rightarrow Q \rightarrow R, \tag{A.2}$$

where P is sufficient and R is necessary for Q.

The relation (A.2) is itself capable of still another interpretation which is also of importance with regard to convergence tests. Let us examine (A.2) relative to R rather than Q. Then we see that both P *and* Q are sufficient for R, but Q is the superior of the two because it is *logically closer* to R: thus the way to improve on a known sufficient condition P for

[1] Cauchy *1821a*, introduction, p. v.

R is to find a new condition Q which interposes itself between the two. Of course, it may not be possible to find such a Q: instead we may have two sufficient conditions P_1 and P_2 for R neither of which implies the other, and the aim will be to move closer to R than at least one of them. In a similar way we try to improve upon known sufficient conditions for R, and so approach R from both sides. We may find a condition S which is both necessary and sufficient for R, in which case the logical position is

$$S \rightarrow R \rightarrow S \tag{A.3}$$

or

$$R \equiv S. \tag{A.4}$$

This is a logical equivalence, which we state as "R if and only if S." It follows that R is also necessary and sufficient for S, and that if we have two such conditions S_1 and S_2 then the situation

$$R \equiv S_1 \equiv S_2 \tag{A.5}$$

holds. Therefore each of the three statements is necessary and sufficient for the other two, and all three are logically equivalent to each other. Cauchy had found such a condition for the convergence of a series: now he was looking for sufficient conditions for convergence which would in practice be easier to implement.

But the situation is a little more complicated than the above logical picture. A test is a statement P from which the convergence of the series Q can be deduced, and it may usually be used also to test for divergence. In general, an expression E is calculated and the test says that if E is less than some numerical value k then the series is convergent, whereas if $E > k$ it is divergent. The situation $E = k$, however, normally means that the test is indecisive. But to talk about the value of E at all is to assume that the expression exists and takes a unique value in the first place; and this amounts to a sufficient condition T on the test itself, by restricting its application to those series for which E can in fact be determined as required. Hence the full logical structure of a test is

$$T \rightarrow (P \rightarrow Q), \tag{A.6}$$

and so the critical improvement on tests involves not only improving on P but on T also. We may seek a condition P for a type of series (or T) which has not previously been investigated; or generalize the types of series (that is, find a logically prior T) to which a particular P applies; or find a new P for a given T for which known P's are indecisive in the way described above.

Cauchy gave the first group of such tests in the *Cours d'analyse* immediately following his disguised assault on Fourier series. By and large, he did not examine the logical relation between tests there, but, on the other hand, he was immediately aware of the distinction between series of terms of the same and of mixed signs. He proved his results for each kind of series separately (where they applied to both), and also stated analogous theorems in the theory of functions where possible. All his expressions E were limiting values, as n tended to infinity, of some function associated with the nth term of the series, and he showed another influence of Bolzano's 1817 paper in always operating with Bolzano's idea of the *upper bound* of the values of a sequence.[2]

Cauchy seems to have come to his tests by comparison of the known convergence of special series. Taking the convergent geometric progression

$$1+x+x^2+\cdots=\frac{1}{1-x}, \qquad |x|<1, \tag{A.7}$$

he remarked that in the series of positive terms

$$u_0+u_1+u_2+\cdots \tag{A.8}$$

the property that

$$0<u_n<x^n<1 \quad \text{when} \quad n>N \tag{A.9}$$

would imply *a fortiori* the convergence of (A.8) from the convergence in (A.7). (A.9) and (A.7) show that

$$0<(u_n)^{1/n}<x<1 \tag{A.10}$$

and so lead to his "root test," which we state in the most general form to include series of mixed terms.

THEOREM A.1 (root test)

If $\overline{\lim}_{n\to\infty}[|u_n|^{1/n}]$ exists and is $\lesssim 1$, then $\Sigma_{r=0}^{\infty} u_r$ is $\substack{\text{convergent}\\ \text{divergent}}$. Otherwise the test is indecisive.[3]

[2] Bolzano *1817a*, art. 12 (quoted in connection with his version of the "Bolzano-Weierstrass theorem" in n. 15, p. 74).

[3] The references to all the tests in the *Cours d'analyse* are given, where applicable, in the order in which they appear there, namely: theory of functions (T), series of the same sign (S), series of mixed sign (M). For the root test, they are: Cauchy *1821a*, pp. 53–54 (T), 132–134 (S), 142–143 (M); *Works*, ser. 2, vol. 3, pp. 58–59 (T), 121–123 (S), 128–129 (M). Cauchy also stated versions of applicable theorems in the field of complex variables in chapter 9 of the *Cours d'analyse*; they have not been included in these footnotes.

The geometrical progression may also have given him the idea for another test. The x in (A.7) is not only equal to $(u_n)^{1/n}$ but also to u_{n+1}/u_n and so we have the "ratio test," approached by d'Alembert:

THEOREM A.2 (ratio test)
If

$$\overline{\lim_{n\to\infty}}\left|\frac{u_{n+1}}{u_n}\right|$$

exists and is $\lessgtr 1$, then $\sum_{r=0}^{\infty} u_r$ is $\begin{smallmatrix}\text{convergent}\\ \text{divergent}\end{smallmatrix}$. Otherwise the test is indecisive.[4]

As Cauchy may have realized, this test is less powerful than the root test since it is also indecisive on all occasions of indecisiveness of the root test.[5] His next test probably came from the series

$$\frac{1}{1^p}+\frac{1}{2^p}+\frac{1}{3^p}+\cdots, \tag{A.11}$$

which was known to be convergent when $p>1$ and divergent when $p\leq 1$. Bolzano had remarked that the divergence when $p=1$ was an example of the necessity, but *not* sufficiency, of the condition for convergence that $u_n\to 0$ as $n\to\infty$ and that its divergence could be proved analytically.[6] Cauchy amplified the point by the (familiar) proof based on the inequality

$$\begin{aligned}&1+\frac{1}{2}+\frac{1}{3}+\frac{1}{4}+\frac{1}{5}+\frac{1}{6}+\frac{1}{7}+\frac{1}{8}+\cdots\\ &>1+\frac{1}{2}+\frac{1}{4}+\frac{1}{4}+\frac{1}{6}+\frac{1}{6}+\frac{1}{8}+\frac{1}{8}+\cdots=\infty,^{7}\end{aligned} \tag{A.12}$$

and then stated

THEOREM A.3
If $u_n>u_{n+1}>0$, then $\sum_{r=0}^{\infty} u_r$ and $\sum_{r=0}^{\infty} 2^r u_{2^r-1}$ converge and diverge together.

[4] Cauchy *1821a*, pp. 48–52 (T), 134–135 (S), 143 (M); *Works*, ser. 2, vol. 3, pp. 54–57 (T), 122–123 (S), 129 (M). Prior to Cauchy, Lacroix was aware of this result, but only from d'Alembert's remarks cited in n. 8 and text, pp. 71–72 (that is, d'Alembert *1768c*). See Lacroix *Treatise*$_1$, vol. 1 (1797), pp. 5–14, and the better treatment in *Treatise*$_2$, vol. 1 (1810), pp. 5–12.
[5] See Cauchy's remark in *1821a*, p. 60; *Works*, ser. 2, vol. 3, pp. 63–64.
[6] Bolzano *1817a*, art. 10.
[7] Cauchy *1821a*, pp. 127–128; *Works*, ser. 2, vol. 3, p. 117.

The proof was based on

$$\begin{aligned} & u_0 + 2u_1 + 2u_2 + 2u_3 + 2u_4 + 2u_5 + 2u_6 + 2u_7 + \cdots \\ & > u_0 + 2u_1 + 2u_3 + 2u_3 + 2u_5 + 2u_5 + 2u_7 + 2u_7 + \cdots \end{aligned} \tag{A.13}$$

in the case of convergence, and

$$\begin{aligned} & u_0 + u_1 + u_2 + u_3 + u_4 + u_5 + u_6 + u_7 + \cdots \\ & < u_0 + u_1 + u_1 + u_3 + u_3 + u_5 + u_5 + u_7 + \cdots \end{aligned} \tag{A.14}$$

in the case of divergence. In fact the convergence of (A.11) could now be proved from this theorem as the mate of the geometric progression;[8] that series gave Cauchy the idea for a new test, which could also work on certain occasions when the ratio test was indecisive. Suppose that we have the condition

$$0 < u_n < \frac{1}{n^p} < 1 \quad \text{when} \quad n > N, \quad p > 1. \tag{A.15}$$

Then $\Sigma_{r=1}^{\infty} 1/r^p$ is convergent, and also

$$\frac{\log(u_n)}{\log(1/n)} = \frac{\log(1/u_n)}{\log(n)} > p > 1. \tag{A.16}$$

Therefore we have

THEOREM A.4 (logarithmic test: for a series of positive terms)
If

$$\overline{\lim_{n\to\infty}} \left[\frac{\log(1/u_n)}{\log(n)}\right]$$

exists and $\gtrless 1$, then $\Sigma_{r=0}^{\infty} u_r$ is $\begin{smallmatrix}\text{convergent}\\\text{divergent}\end{smallmatrix}$ Otherwise the test is indecisive.[9]

Now Cauchy moved to tests involving a series made up of two given series. The most interesting result was

THEOREM A.5
Let $\Sigma_{r=0}^{\infty} u_r$ and $\Sigma_{r=0}^{\infty} v_r$ be two absolutely convergent series with sums s and s'. Then $\Sigma_{n=0}^{\infty} w_n$, where w_n is the "Cauchy product"

$$w_n = \sum_{t=0}^{n} u_t v_{n-t},$$

is convergent to the sum ss'.[10]

[8] Cauchy *1821a*, pp. 135–137; *Works*, ser. 2, vol. 3, pp. 123–125.
[9] Cauchy *1821a*, pp. 137–140; *Works*, ser. 2, vol. 3, pp. 125–127.
[10] Cauchy *1821a*, pp. 141–142 (S), 147–150 (M); *Works*, ser. 2, vol. 3, pp. 127–128 (S), 132–135 (M).

Finally he gave the "alternating series test," familiar from the seventeenth century as a means of estimating sums of series to which it is applicable:

THEOREM A.6 (alternating series test)
If u_n alternates in sign with successive values of n and tends to zero as $n \to \infty$, then $\Sigma_{r=0}^{\infty} u_r$ is convergent.[11]

Cauchy was now ready to apply these tests to analytical problems. The first was concerned with power series, where he discovered from the root test that $\Sigma_{r=0}^{\infty} a_r x^r$ was $\substack{\text{convergent}\\\text{divergent}}$ for all values of x for which

$$|x| \lessgtr \left[\overline{\lim_{n\to\infty}}[|a_n|^{1/n}]\right]^{-1}, \tag{A.17}$$

(leaving to Abel the investigation of the situation at the critical value of x).[12] Another property which he announced was

THEOREM A.7
"A continuous function of the variable x can only be developed in a unique manner in a convergent series, ordered according to increasing and whole powers of that variable."[13]

Both (A.17) and theorem A.7 have obvious implications for Taylor's series

$$f(x+h) = f(x) + hf'(x) + \frac{h^2}{2!}f''(x) + \cdots. \tag{A.18}$$

The inequalities (A.17) tell us the range of convergence of the series, and the theorem implies that for suitable functions the series will be the Taylor series. Cauchy dealt with the series in the *Résumé*: having obtained Lagrange's remainder forms

[11] Cauchy *1821a*, pp. 144–145; *Works*, ser. 2, vol. 3, pp. 130–131.
[12] Cauchy *1821a*, pp. 151; *Works*, ser. 2, vol. 3, p. 136. See also the (less general) application of the ratio test to power series immediately following this result.
[13] Cauchy *1821a*, pp. 162–164; *Works*, ser. 2, vol. 3, pp. 144–145. Cauchy's proof was based on taking two equal power series and successively proving the equality of corresponding coefficients. Abel either missed this theorem or was not satisfied with the proof, for he proposed the same problem in the form:

$$\text{If} \quad \sum_{r=0}^{\infty} a_r x^r = 0, \quad \text{then all } a_n = 0, \tag{1}$$

to Crelle in a letter of 1827 (see Abel *Correspondence*, pp. 63–64). Crelle published it along with other problems in his journal (see Abel *1827a*).

$$\frac{1}{(n-1)!}\int_0^h (h-t)^{n-1} f^{(n)}(x+t)\,dt \tag{A.19}$$

and

$$\frac{1}{n!}h^n f^{(n)}(x+\theta h), \qquad 0<\theta<1, \tag{A.20}$$

for him,[14] he then showed that the reverse kind of result is not true with the counterexample to it of $\exp(-1/x^2)$, which is zero along with all its derivatives when $x=0$ and so does not take the Taylor series around that value. Hence for any function $f(x)$ which does have a convergent Taylor series around $x=0$, the function $[f(x)+\exp(-1/x^2)]$ has the same one, which refutes the converse of Cauchy's theorem A.7—and also destroys the last remnants of Lagrange's foundations of the calculus.[15]

The next batch of tests came from Abel. In his 1826 paper in Crelle's journal on the binomial series he introduced tests for series with terms of which the binomial

$$\frac{m(m-1)\cdot\cdots\cdot(m-r+1)}{1\cdot 2\cdot\cdots\cdot r}x^{m-r}$$

is an example: where the term can be taken as the product of two other terms. His first result was a version of "Abel's test":

THEOREM A.8
If $\Sigma_{r=0}^{\infty} u_r$ is a convergent series of positive terms and $0<v_n<1$, $n=1, 2, \ldots$, then $\Sigma_{r=0}^{\infty} u_r v_r$ is convergent.[16]

[14] Cauchy *1823a*, lecture 36, equations (3)–(12). Regarding (A.20), see also Cauchy *1826b*, and for Lagrange's work on these forms, see n. 2 and text, p. 68. Bolzano's 1830s manuscript contains a long discussion of the convergence of Taylor's series and the remainder form (A.19), in both one and several variables (Bolzano *Writings*, vol. 1, pp. 155–183).

[15] Cauchy *1823a*, lecture 38: he first made the point in *1822a*. See also further investigations of the series and its remainder in the "Addition" and in the section "Sur les séries de Taylor et MacLaurin" at the end of the book. The *Leçons* (*1829a*) appear to contain no new results: the main passages are in lectures 8–10.

[16] Abel *1826b*, theorem II: theorem I is the corresponding divergence test. Abel did not say that $\Sigma_{r=0}^{\infty} u_r$ was convergent but stated the condition from which convergence follows by the ratio test. The result, usually known today as "Abel's test," may be stated: If $\Sigma_{r=0}^{\infty} u_r$ is a convergent series and $v_1, v_2, \ldots$ a monotonic convergent sequence, then $\Sigma_{r=0}^{\infty} u_r v_r$ is convergent. The name "Abel's test" seems to be due to Jordan, whose *Cours d'analyse* was published in three editions between the 1880s and the 1910s and was the most important of the series of works with this title after Cauchy's. (See Jordan *Course*$_1$, vol. 1 (1882), art. 125; *Course*$_2$, vol. 1 (1893), art. 302; *Course*$_3$, vol. 1 (1909), art. 302.)

The next result was

THEOREM A.9
If the nth partial sum of $\Sigma_{r=0}^{\infty} u_r$ is bounded above by the value δ, and the sequence $v_1, v_2, \ldots$ of positive terms decreases monotonically, then

$$u_0 v_0 + u_1 v_1 + \cdots + u_n v_n < u_0 \delta.^{17} \tag{A.21}$$

It was to prove this theorem that Abel devised his "partial summation formula":

$$\begin{aligned} &u_0 v_0 + u_1 v_1 + \cdots + u_n v_n \\ &= (u_0 - u_1)s_0 + (u_1 - u_2)s_1 + \cdots + (u_{n-1} - u_n)s_{n-1} + u_n s_n, \end{aligned} \tag{A.22}$$

where

$$s_r = \sum_{j=0}^{r-1} u_j, \tag{A.23}$$

and he used the theorem in the noncontroversial parts of his proof of his limit theorem (theorem 4.5). But beyond that it has a generalization to a full test which manifested itself in Dirichlet's use of the partial summation formula to give his own proof of the limit theorem (in the form of theorem 5.4), and so is now known after him:

THEOREM A.10 (Dirichlet's test)
If $\Sigma_{r=0}^{\infty} u_r$ is a series with bounded partial sums and $v_1, v_2, \ldots$ decreases monotonically to zero, then $\Sigma_{r=0}^{\infty} u_r v_r$ is convergent.[18]

Finally Abel gave a new proof of Cauchy's theorem A.5 on the product of two absolutely convergent series. He was not able to relax its conditions, but he did give a new proof based on the method of his limit theorem, where $\Sigma_{r=0}^{\infty} u_r$ and $\Sigma_{r=0}^{\infty} v_r$ were taken as the limiting values of $\Sigma_{r=0}^{\infty} u_r \alpha^r$ and $\Sigma_{r=0}^{\infty} v_r \alpha^r$ as $\alpha \to 1$.[19]

In the next volume of Crelle's journal there appeared a paper from a certain Louis Olivier. Olivier failed to understand the logic underlying the development of tests, and instead of trying to extend known results

[17] Abel *1826b*, theorem III.
[18] The test seems to be due to Jordan (see Jordan *Course*$_2$, vol. 1 (1893), art. 299; *Course*$_3$, vol. 1 (1909), art. 299). For Dirichlet's proof of Abel's limit theorem see n. 27 and text, p. 109. The u_n of theorem A.10 correspond to the v_n there, and the v_n to the α^n, $\alpha < 1$.
[19] Abel *1826b*, theorem VI. Not until the 1870s were the conditions on this theorem relaxed: then F. Mertens found that absolute convergence need be assumed of only one of the two series (see Mertens *1875a*).

or introduce another one into the discussion, he produced his "general criterion" for distinguishing between convergence and divergence:

THEOREM A.11

If $nu_n \overset{\to}{\nrightarrow} 0$ as $n \to \infty$ then $\Sigma_{r=0}^{\infty} u_r$ is $\substack{\text{convergent} \\ \text{divergent}}$. The result is also true if we replace u_n by the nth group of terms with the same sign.[20]

Not only was the thinking behind the "general criterion" muddled, but even as a modest test it was faulty. The proof (which we give in the case of individual terms) was based on the inequality

$$nu_n > r_{n-1}, \tag{A.24}$$

where r_{n-1} is the remainder of the series after n terms: if $nu_n \to 0$ as $n \to \infty$, then $r_{n-1} \to 0$ also, and therefore the series is convergent. However, the further inequality

$$nu_n > r_{n-1} > nu_{2n} \tag{A.25}$$

shows that if $nu_n \nrightarrow 0$ as $n \to \infty$, then $r_n \nrightarrow 0$ either, and so the series is divergent.[21]

The divergence result is true; but the proof of convergence is inaccurate. Neither (A.24) nor (A.25) follow from their respective assumptions, and Abel pointed out a counterexample in a note of the following year (1828):

$$u_{n-2} > \frac{1}{n \log n}, \qquad n = 2, 3, \ldots, \tag{A.26}$$

which he showed to give a divergent series, although it satisfied Olivier's criterion.[22] Olivier accepted the criticism in an appendix to Abel's note,

[20] Olivier *1827a*, p. 34. In Crelle's journal Olivier is said to be "at that time at Berlin": he published several small papers in its first three volumes. No other information on him seems to be available: he was probably not the French mathematician Theodore Olivier (1793–1853), who wrote mainly on differential geometry.

[21] Olivier *1827a*, pp. 31–34: see also the examples on pp. 34–42.

[22] Abel *1828a*. His demonstration of divergence followed from the inequality

$$\log \log(1 + n) < \log \log 2 + \sum_{r=2}^{n} \frac{1}{r \log r}. \tag{1}$$

It is this counterexample which Pringsheim later used to refute the "Euler criterion" for convergence (see n. 18, p. 75). A manuscript, first published in the Sylow-Lie edition of Abel's Collected Works and dated by the editors from the second half of 1825 onward, contains draft criticisms of Olivier by means of this counterexample (Abel *Series*; see *Works*$_2$, vol. 2, pp. 199–201). About 25 years after this discussion a certain J. Tetmayer, presumably ignorant of it, announced Olivier's criterion as his general rule for convergence (see Tetmayer *1851a*). And at the end of the century, the criterion appeared as a condition for the validity of the *converse* of Abel's limit theorem (theorem 4.5). (See Tauber *1897a*.)

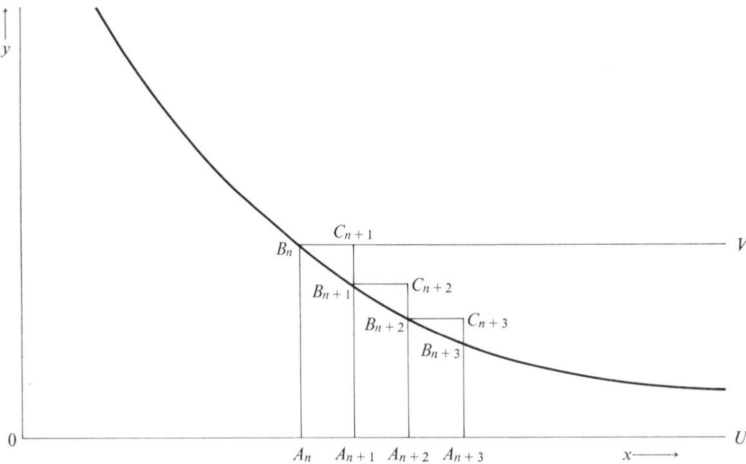

DIAGRAM A.1

and in showing the line of thought which had led him to the test, he partially explained and completely revealed its weakness. He had made use of the area interpretation of the integral: in his diagram (relettered here as diagram A.1), u_n is given by the ordinate $A_n B_n$ of the curve for the abscissa $OA_n = n$ and so the terms have to obey the restriction—not stated in Olivier's paper—of being positive and tending monotonically to zero. His reasoning had been based on the inequality

$$\square(A_n B_n VU) > \square(A_n B_n C_{n+1} A_{n+1}) + \square(A_{n+1} B_{n+1} C_{n+2} A_{n+2}) + \cdots \tag{A.27}$$

which leads to

$$\begin{aligned} nu_n &> 1 \cdot u_n + 1 \cdot u_{n+1} + \cdots \\ &= r_{n-1}. \end{aligned} \tag{A.28}$$

But it is $\square(A_n B_n C_{2n-1} A_{2n-1})$, and not $\square(A_n B_n VU)$, which equals nu_n, and although he could have effected some repair by assigning U to $x = 2n - 1$ and deducing that

$$nu_n > u_n + u_{n+1} + \cdots + u_{2n-1}, \tag{A.29}$$

(A.24) itself still does not follow. As Olivier reminded himself, arguments for finite situations are not necessarily valid in the infinite case, and so he suggested a revised convergence test based on the successive application of (A.24) for n, $2n$, ... to deduce that

$$n(u_n + u_{2n} + \cdots) > u_n + \cdots + u_{2n} + \cdots \tag{A.30}$$

and hence:

THEOREM A.12
If $n(u_n + u_{2n} + \cdots) \to 0$ as $n \to \infty$, so does r_n, and if $nu_n \to 0$ as $n \to \infty$, then $\Sigma_{r=0}^{\infty} u_r$ is convergent.[23]

But this result is no improvement over its predecessor if the same restrictions on the u_r still apply, and it is not necessarily true (because (A.25) is not necessarily true) if they do not. Olivier's mistake and Abel's criticism are important both as a reminder that the story is not always one of serene success, and also to form the background to the successful part of Olivier's original 1827 paper, which seems to have come to him as a corollary of his general theorem A.11:

THEOREM A.13
If $u_1, u_2, \ldots$ is a sequence of positive terms which decrease monotonically to zero as $n \to \infty$ when $n > m$, then, by theorem A.11, $\Sigma_{r=0}^{\infty} u_r$ is convergent. Let us call the sum of the series s, and call $u(x)$ a monotonically decreasing function which equals u_n when $x = n$. Then

$$0 < \sum_{r=m+1}^{\infty} u_r < \int_m^{\infty} u(x)\, dx < \sum_{r=m}^{\infty} u_r \tag{A.31}$$

and

$$\int_m^{\infty} u(x)\, dx < s < \int_m^{\infty} u(x)\, dx + u_m.^{24} \tag{A.32}$$

The proof of (A.31) and (A.32) follow immediately (and accurately) from diagram A.2:

$$\begin{aligned} &\square(A_m D_m B_{m+1} D_{m+1} B_{m+2} D_{m+2} \cdots) < \int_m^{\infty} u(x)\, dx \\ &< \square(A_m B_m C_{m+1} B_{m+1} C_{m+2} B_{m+2} \cdots). \end{aligned} \tag{A.33}$$

The point of the theorem for Olivier was to estimate the sum of a series

[23] Olivier *1828a*.
[24] Olivier *1827a*, pp. 42–44.

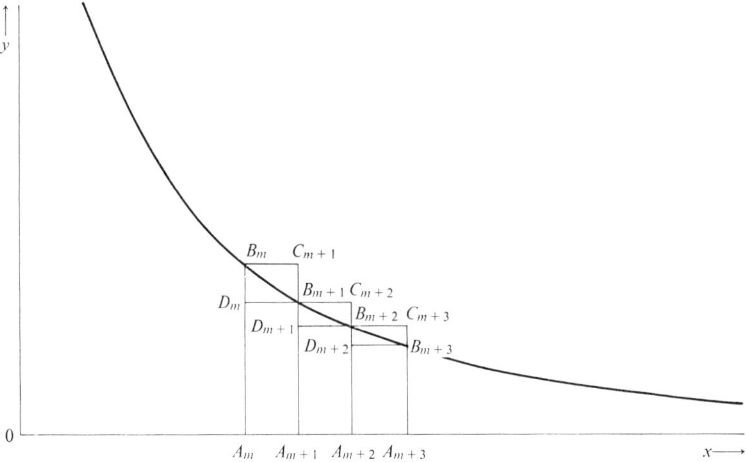

DIAGRAM A.2

whose convergence would have been already established by his theorem A.11. As an estimator it had already been known in the eighteenth century to Euler and to Colin MacLaurin (1698–1746);[25] and if we abandon Olivier's theorem A.11 (as we must!) we see that theorem A.13 may itself be taken as a test for convergence and divergence. This is precisely what Cauchy did in a paper published in his own journal later in the year 1827 of Olivier's paper:

THEOREM A.14 (Cauchy's integral test)
Let $u_n > 0$ and decrease monotonically to zero as $n \to \infty$, when $n \geq m$. Then $\Sigma_{r=m}^{\infty} u_r$ and $\int_m^{\infty} u(x)\,dx$ converge or diverge together.[26]

Olivier is not mentioned, and in fact it is quite possible that Cauchy came to his result independently, for he presented it as one means of solving the cases of indecision of the root test (theorem A.1):[27] a point

[25] Euler *1736a*; and MacLaurin *1742a*, art. 350 (which in *1894a* Eneström surprisingly saw as a general anticipation of convergence). Later Euler showed that he knew how to use the double integral as a volume (see Euler *1769a*, esp. art. 8).
[26] Cauchy *1827b*, theorem 2: see also a reformulation of the necessary and sufficient condition for convergence in theorem 1. To state the monotonic decrease of the u_n Cauchy wrote that it "decreases without stopping with $1/x$," which sounds as if the order of decrease must be that of $1/x$; but his proof makes clear that he intended the more general interpretation that $f(x)$ decreases monotonically, as does $1/x$.
[27] Cauchy *1827b*, p. 222; *Works*, ser. 2, vol. 7, p. 268.

which epitomized the new period of tests which it inaugurated. For now that Cauchy and Abel had produced a series of "basic" tests, the task was to build upon them by finding new tests which dealt with series not yet covered or where existing tests were indecisive. Joseph Raabe (1801–1859) was the next to write in this vein: like Abel and Olivier, he acknowledged Cauchy's *Cours d'analyse* as his starting point, but he did not seem to be aware of Cauchy's 1827 paper, for his first result was Cauchy's integral test.[28] From there he proceeded to further results of which the following was the most general and is now known after him:

THEOREM A.15 (Raabe's test)
If $u_r > 0$ and

$$\lim_{n\to\infty}\left[n\left(\frac{u_n}{u_{n+1}}-1\right)\right]$$

exists and is $\gtrless 1$, then $\Sigma_{r=0}^{\infty} u_r$ is $\substack{\text{convergent}\\\text{divergent}}$. Otherwise the test is indecisive.[29]

Raabe used his new tests on the hypergeometric series

$$1+\frac{\alpha\cdot\beta}{1\cdot\gamma}x+\frac{\alpha(\alpha+1)\cdot\beta(\beta+1)}{1\cdot 2\cdot\gamma(\gamma+1)}x^2+\cdots. \tag{A.34}$$

When $x=1$ in (A.34), u_n/u_{n+1} takes the form

$$\frac{u_n}{u_{n+1}}=\frac{n^\lambda+q_1n^{\lambda-1}+\cdots}{n^\lambda+p_1n^{\lambda-1}+\cdots} \tag{A.35}$$

$$=1+\frac{(q_1-p_1)n^{\lambda-1}+(q_2-p_2)n^{\lambda-2}+\cdots}{n^\lambda+p_1n^{\lambda-1}+\cdots}. \tag{A.36}$$

Therefore, relative to his theorem A.15,

$$n\left(\frac{u_n}{u_{n+1}}-1\right)=\frac{(q_1-p_1)+(q_2-p_2)n^{-1}+\cdots}{1+p_1n^{-1}+\cdots} \tag{A.37}$$

and so

$$\left|n\left(\frac{u_n}{u_{n+1}}-1\right)\right|\to|q_1-p_1| \quad \text{as} \quad n\to\infty. \tag{A.38}$$

[28] Raabe *1832a*, arts. 1–6, including examples of its use.
[29] Raabe *1832a*, art. 11. See also a test in art. 8 based on the expression

$$\lim_{n\to\infty}\left[n\log\left(\frac{u_n}{u_{n+1}}\right)\right]. \tag{1}$$

A review of the "state of the art" was given in Klügel *1833b*, and also in part in Klügel *1833a*, pp. 283–289.

Now the test showed that the series was convergent or divergent according as $|q_1 - p_1|$ was greater or less than 1, confirming Gauss's own analysis of the series.[30]

But to what work of Gauss was Raabe referring? In 1813 Gauss had published a paper giving a general examination of the hypergeometric series, which had arisen for him in connection with certain problems in theoretical astronomy.[31] One of the properties which he investigated was convergence, which, as we saw earlier, he understood as a general problem at that time along with so few others.[32] His approach had been to transform the series, which we may denote by $\Sigma_{r=0}^{\infty} u_r$, into $\Sigma_{r=0}^{\infty} v_r$ by relations such as

$$v_n = u_n \left(\frac{n-1}{n}\right)^h; \tag{A.39}$$

and to compare $\Sigma_{r=0}^{\infty} v_r$ with series of known convergence or divergence. The test which resulted was

THEOREM A.16

If u_n/u_{n+1} is expressed in the form (A.35) and $|q_1 - p_1|$ is $\lesseqgtr 1$, then $\Sigma_{r=0}^{\infty} u_r$ is $\begin{smallmatrix}\text{convergent}\\ \text{divergent}\end{smallmatrix}$.[33]

We see at once what Raabe may not have noticed: if the limit in (A.38) is greater or less than 1, then the two tests are coincidental. But Gauss's test goes further, to show the divergence of the hypergeometric series when the limit is exactly 1 under the conditions of the structure of (A.35). Gauss was indeed the master; and by so much was he the master that twenty years had elapsed since the publication of his paper before anybody was able to catch up with him.[34] In 1813 hardly anyone else understood what the convergence problem was: yet in his aloof way, Gauss concentrated only on the further, technical question of the convergence of the particular series in which he was interested, and one can readily assert that Raabe himself only appreciated the purpose of Gauss's reasoning

[30] Raabe *1832a*, art. 12. In art. 9, he applied his above test of n. 29 to the hypergeometric series.

[31] In fact, with the coefficients in the expansion of $(a^2 + b^2 - 2ab\cos\phi)^{-n}$ as a trigonometric series (see Gauss *1813a*, art. 6).

[32] See n. 9 and text, p. 71.

[33] Gauss *1813a*, art. 16, esp. sects. IV and V for divergence and convergence respectively. In the undated manuscript *Equation* Gauss took his examination of the series further, although not from the point of view of convergence. For commentary on the paper, see Schlesinger *Gauss*$_1$ (1912), esp. pp. 95–100; and *Gauss*$_2$ (1933), pp. 134–155.

[34] An earlier mention of Gauss's result is to be found in Klügel *1823a*, pp. 589–591.

because he had read Cauchy's textbook first. The result, which is now known after Gauss, is not in the form of theorem A.16, but a generalization based on (A.36) and therefore close to Raabe's test:

THEOREM A.17 (Gauss's ratio test)
If u_n/u_{n+1} can be expressed in the form

$$\frac{u_n}{u_{n+1}} = 1 + \frac{\mu}{n} + \frac{K_n}{n^{1+p}}, \tag{A.40}$$

where $|K_n|$ is bounded and $p > 0$, and if μ is $\gtrless 1$, then $\Sigma_{r=0}^{\infty} u_r$ is $\substack{\text{convergent} \\ \text{divergent}}$.[35]

Shortly after the appearance of Raabe's paper another German mathematician sent in to Crelle's journal a paper on convergence, which was published in 1835. Ernst Kummer (1810–1893) may not have known of Raabe's paper, which appeared in another journal (although Raabe did publish a summary in Crelle's journal in 1834, after Kummer's paper had been written but before it was published);[36] but he certainly knew of Gauss's paper and the Olivier-Abel discussion and his main test was a generalization of Raabe's test and Olivier's theorem A.11:

THEOREM A.18
If $u_n > 0$ when $n > N$ and $m_1, m_2, \ldots$ is a sequence of integers such that $\lim_{n\to\infty}[m_n u_n] = 0$ and $\lim_{n\to\infty}\left[\frac{m_n u_n}{u_{n+1}} - m_{n+1}\right]$ exists and is >0, then $\Sigma_{r=0}^{\infty} u_r$ is convergent. If this latter limit is zero and

$$\lim_{n\to\infty}\left[\frac{m_n u_n}{\dfrac{m_n u_n}{u_{n+1}} - m_{n+1}}\right] \neq 0,$$

then $\Sigma_{r=0}^{\infty} u_r$ is divergent.[37]

[35] The test seems to have been introduced in this form by K. Knopp (see Knopp *Series*$_1$, (1928), pp. 288–289; *Series*$_2$ (1951), pp. 288–289), although it is implied in Weierstrass's investigations of the form

$$\frac{u_n}{u_{n+1}} = 1 + \frac{a_1}{n} + \frac{a_2}{n^2} + \frac{a_3}{n^3} + \cdots \tag{1}$$

(See Weierstrass *1856a*, pp. 22-25; *Works*, vol. 1, pp. 177–181.)

[36] Raabe *1834a*.

[37] Kummer *1835a*, pp. 172 and 173. The theorem is one of great generality, and has been taken to be some kind of "all-embracing" test (which, of course, the logic of tests shows that it cannot be). (See especially du Bois Reymond *1872a*. For descriptive commentary see Cajori *1893a*, pp. 6–10.)

The results of Olivier and Raabe follow from the special case of

$$m_n = n, \tag{A.41}$$

when the first condition for convergence becomes Olivier's

$$\lim_{n\to\infty} [nu_n] = 0 \tag{A.42}$$

and the second Raabe's

$$\lim_{n\to\infty} \left[n\left(\frac{u_n}{u_{n+1}} - 1\right) - 1\right] = 0. \tag{A.43}$$

Kummer drew these conclusions explicitly in his paper: he saw the frailty of Olivier's result by using the divergence part of his own theorem to prove the divergence of Abel's counterexample (A.26) to it,[38] and stated Raabe's test explicitly as a deduction from (A.43).[39]

The next author was Jean Duhamel (1797–1872). Duhamel seems to have read nothing on convergence beyond Cauchy's *Cours d'analyse*, for in 1839 he laid claim both to the integral test and to Raabe's test. This provoked an incident which showed that the standards of personal attack in Paris were not falling. In a later volume of the journal in which Duhamel's paper appeared, Victor Lebesgue (1791–1875) published a French translation of Raabe's 1834 summary paper from Crelle's journal, with a footnote which implied that Duhamel's ignorance of Raabe's work was due to his ignorance of German, and announcing that "the persons who have read Mr. Duhamel's article will see with pleasure the extension given by Mr. Raabe to the rule in question."[40] But Duhamel's formulation of the test seems to be logically equivalent to Raabe's; and he also announced the generalized comparison test which many of his predecessors had been using in particular cases:

[38] Kummer *1835a*, pp. 175–176; see also pp. 178 and 180.

[39] Kummer *1835a*, p. 177. On p. 178 he followed Raabe in examining the hypergeometric series (A.34), and in *1836a* he followed Gauss in devoting a special paper to its properties. On pp. 180–181 of *1835a* he stated a test for $\sum_{r=0}^{\infty} u_r v_r$ based on the convergence of the sequence $u_1, u_2, \ldots$ and the series $\sum_{r=0}^{\infty} v_r$, which was similar to the (later) Dirichlet's test (theorem A.10), and on p. 183 the mutual convergence and divergence of $\sum_{r=0}^{\infty} a_r$ and $\prod_{r=0}^{\infty} (1 + a_r)$. (On infinite products, compare Cauchy *1821a*, note IX.)

[40] Raabe *1841a*, p. 85. For Duhamel's theorems, see his *1839a*, pp. 215–219 and 219–221.

THEOREM A.19 (comparison test)
If $\Sigma_{r=0}^{\infty} u_r$ is $\begin{smallmatrix}\text{convergent}\\ \text{divergent}\end{smallmatrix}$ and

$$\frac{u_{n+1}}{u_n} \gtrless \frac{v_{n+1}}{v_n} \qquad \text{when} \quad n \geq N, \tag{A.44}$$

then $\Sigma_{r=0}^{\infty} v_r$ is also $\begin{smallmatrix}\text{convergent}\\ \text{divergent}\end{smallmatrix}$.[41]

After Duhamel the development of tests entered a more sophisticated phase, inaugurated by Augustus de Morgan (1806–1871), the first English mathematician of the time to make a significant contribution to mathematical analysis. De Morgan's understanding of convergence was rather shaky: he talked in d'Alembertian terms of series which "begin divergently"[42] and shared with Poisson an inability to appreciate the significance of Abel's limit theorem and so defended Poisson's views on convergence.[43] But in an installment of 1839 of his treatise on the calculus he did introduce a test which contained the important novelty of its own—*a method of iteration* in cases of indecision:

THEOREM A.20
Let $u_n > 0$ and the sequence $u_1, u_2, \ldots$ decrease monotonically to zero, and let

$$p_0 = -\frac{nu_n'}{u_n}, \qquad u_n' \equiv \frac{du_n}{dn}. \tag{A.45}$$

Then if $\lim_{n\to\infty}[p_0]$ exists and is $\gtrless 1$, $\Sigma_{r=0}^{\infty} u_r$ is $\begin{smallmatrix}\text{convergent}\\ \text{divergent}\end{smallmatrix}$. If the limit is 1 the test is indecisive, and we repeat it successively on $p_1, p_2, p_3, \ldots$, where

$$\begin{aligned} p_1 &= \log[n(p_0 - 1)] \\ p_2 &= \log\log[n(p_1 - 1)] \\ p_3 &= \log\log\log[n(p_2 - 1)] \end{aligned} \tag{A.46}$$

....[44]

De Morgan's idea of iterated tests was taken up by a young French student called Joseph Bertrand (1822–1900). Bertrand made ingenious use

[41] Duhamel *1839a*, pp. 214–215.
[42] De Morgan *1842a*, p. 222. For d'Alembert's ideas on convergence, see n. 8 and text, pp. 70–71.
[43] See de Morgan *1842a*, ch. 19, and parts of ch. 20; *1844a*, pp. 182–187; and *1864a*. For commentary, see Burkhardt *1911a*, pp. 187–190.
[44] De Morgan *1842a*, p. 326; he wrote $1/\phi(x)$ for our u_n. On the previous page he gave the integral test and on pp. 235–236 a comparison test anticipated in form by Cauchy in the paper where he had introduced the integral test (see Cauchy *1827b*, theorem 3). We have cited de Morgan's treatise as *1842a*: it appeared in 25 installments between 1836 and 1842, with the dates given on p. iv.

of known tests to develop an iteration on Cauchy's logarithmic test with the expressions

$$\lim_{n\to\infty}\left[\frac{\log\dfrac{1}{u_n}}{\log n}\right],\quad \lim_{n\to\infty}\left[\frac{\log\dfrac{1}{nu_n}}{\log\log n}\right],\quad \lim_{n\to\infty}\left[\frac{\log\dfrac{1}{nu_n\log n}}{\log\log\log n}\right],\ldots \tag{A.47}$$

(where the value of indecision on each limit is unity), and also on Raabe's test, where he wrote u_{n+1}/u_n in the forms

$$\frac{1}{1+\alpha},\quad \frac{1}{1+\dfrac{1}{n}+\alpha'},\quad \frac{1}{1+\dfrac{1}{n}+\dfrac{1}{n\log n}+\alpha''},\ldots \tag{A.48}$$

inspired by Duhamel's version of the test in his 1839 paper, and tested the expressions

$$\lim_{n\to\infty}[n\alpha],\quad \lim_{n\to\infty}[n\alpha'\log n],\quad \lim_{n\to\infty}[n\alpha''\log n\log\log n],\ldots \tag{A.49}$$

for their values relative to unity.[45] Then in a logical survey he showed that the logarithmic test, Raabe's test and de Morgan's theorem A.20, and also their respective iterations, were logically equivalent to each other.[46]

[45] Bertrand *1842a*, pp. 37–48. Both proofs were based on Duhamel's comparison theorem A.19. Compare Abel's manuscript *Series*, p. 201.

[46] Bertrand *1842a*, pp. 48–51. The equivalence of the logarithmic test and de Morgan's theorem A.20 followed from

$$\lim_{n\to\infty}\left[\frac{\log\dfrac{1}{u_n}}{\log n}\right]=\lim_{n\to\infty}\left[\frac{\dfrac{d}{dn}\left(\log\dfrac{1}{u_n}\right)}{\dfrac{d}{dn}(\log n)}\right]=\lim_{n\to\infty}\left[-\frac{nu_n'}{u_n}\right]. \tag{1}$$

For Raabe's test and de Morgan's theorem A.20, Bertrand showed that

$$n\alpha=-\frac{n}{u_n}u'(X),\qquad n\leq X\leq n+1, \tag{2}$$

where $u(x)=u_n$ when $x=n$, giving

$$n\alpha=\left[-\frac{nu_n'}{u_n}\right]\left[\frac{u'(X)}{u_n'}\right]. \tag{3}$$

Therefore

$$\lim_{n\to\infty}[n\alpha]=\lim_{n\to\infty}\left[-\frac{nu_n'}{u_n}\right], \tag{4}$$

from which the equivalence follows. Like Kummer, Bertrand then treated the hypergeometric series (A.34) and referred to Gauss's paper *1813a*. However, he did not mention Kummer *1835a*, although he must surely have known of it. Twenty years later he did include Gauss's theorem A.16, Kummer's theorem A.18, and his own iterations in his treatise on the calculus (see Bertrand *Calculus*, vol. 1 (1864), arts. 237–248).

In the following year Ossian Bonnet (1819–1892) produced a similar paper on iterations. Indeed, some of his results were similar to Bertrand's; Bonnet's aggrieved implication was that his results had been developed independently. His principal new result was an iteration on the root test with the expressions:

$$\lim_{n\to\infty}[|u_n|^{1/n}],\quad \lim_{n\to\infty}\left[\frac{n(1-|u_n|^{1/n})}{\log n}\right],\quad \lim_{n\to\infty}\left[\frac{n(1-n^{-1}\log n-|u_n|^{1/n})}{\log\log n}\right],\ldots \tag{A.50}$$

with unity again the critical value.[47] Another iteration on this test came in 1851 from Magnus von Paucker (1787–1855): he took the logarithm of Cauchy's $|u_n|^{1/n}$ and so tested successively

$$\lim_{n\to\infty}\left[\frac{\log\dfrac{1}{u_n}}{n}\right],\quad \lim_{n\to\infty}\left[\frac{\log\dfrac{1}{nu_n}}{\log n}\right],\quad \lim_{n\to\infty}\left[\frac{\log\dfrac{1}{nu_n\log n}}{\log\log n}\right],\ldots \tag{A.51}$$

relative to zero. He also suggested an interesting result in which Raabe's test was the first iteration on the ratio test, by examining the limiting values of $q_0, q_1, q_2, \ldots$ relative to zero, where

$$\begin{aligned}
q_0 &= \frac{u_n}{u_{n+1}} - 1 \text{ (for the ratio test)}\\
q_1 &= q_0 n - 1 \text{ (for Raabe's test)}\\
q_2 &= q_1 \log n - 1\\
q_3 &= q_2 \log\log n - 1\\
&\ldots.
\end{aligned} \tag{A.52}$$

Finally he showed that his two iterations were logically equivalent.[48]

Iterative tests soon fell into obscurity, presumably because of the complicated expressions involved in the iterations. The results on the equivalence of certain tests, however, were effective in eliminating superfluous theorems: Raabe's test won the battle with Cauchy's logarithmic

[47] Bonnet *1843a*, pp. 99–106: like Bertrand he also used comparison methods in his proofs. He assumed that each $u_n > 0$, but our use of $|u_n|$ to cover negative terms brings the iteration fully in line with Cauchy's original result.

[48] Von Paucker *1851a*, pp. 139–150. His proof of both iterations used Cauchy's theorem A.3 on the mutual convergence and divergence of $\sum_{r=0}^{\infty} u_r$ and $\sum_{r=0}^{\infty} 2^r u_{2^r-1}$, and of their equivalence Cauchy's function-theoretic version of the ratio test (see n. 4, p. 135):

$$\lim_{x\to\infty}[f(x+1)-f(x)] = \lim_{x\to\infty}\left[\frac{f(x)}{x}\right]. \tag{1}$$

test and de Morgan's test because of the comparative ease of the calculation of its expression. But along with redundant tests has gone the *logical intensity of development* which characterized the discovery of all the results: from the first appearance of books on convergence and treatises on analysis with substantial convergence sections, tests were presented as a largely unconnected list.[49] We do not always take only the best from our predecessors' work.

[49] After two unimportant books on convergence (Catalan *1860a* and Laurent *1862a*), the most significant works to contain detailed examinations of convergence substantially beyond Cauchy's *Cours d'analyse* were Bertrand's *Calculus*, already mentioned in n. 46, p. 149 (see vol. 1 (1864), arts 229–271), and Jordan's *Course*$_1$, mentioned in n. 16, p. 138 (see vol. 1 (1882), arts. 109–129). The distinction between "absolute" and "conditional" convergence was further developed (see especially du Bois Reymond *1870a* and Pringsheim *1897a*), although Cauchy had made use of it in the *Cours d'analyse* and later work in developing the theory of convergence of double series (see Cauchy *1821a*, note VII; *1833a*, art. 8; and *1844c*). With the advent of modes of convergence of series of functions, various tests—especially Abel's and Dirichlet's (theorems A.8 and A.10)—could be extended to cover uniform convergence. New tests were also stated, of which the most important was "Weierstrass's test":

If $|u_n(x)| < \delta_n$ when $a \leq x \leq b$ for all n, and $\sum_{r=0}^{\infty} \delta_r$ converges, then $\sum_{r=0}^{\infty} u_r(x)$ converges uniformly over $[a, b]$. (Weierstrass *1880a*, pp. 719–720 (footnote); *Works*, vol. 2, p. 202 (footnote).)

BIBLIOGRAPHY

For economy of presentation of the bibliography, only the first phrase of a long title has been given, titles of journals have been reduced to the abbreviation of key words, and only the principal reissues, reprints and translations of our acquaintance have been mentioned. In addition the words "series," "volume," and "pages" have been omitted in references to journals, so that, for example, "series 2, volume 11 (1819: published 1822), pages 117–125." becomes "(2)11(1819: publ. 1822), 117–125." Explanatory remarks of our own have been inserted in square brackets.

The works of each author are presented in the order: Collected, or collections of, Works; dated references, in chronological order; catchwords, in alphabetical order. We recall that the dating reference to a paper is normally to the designated year of the journal concerned; but when the journal was assigned to a period of years or our text has required mention of the (different) date of composition or presentation, then that year has been used instead.

Abel, N. H.

Works$_1$. *Oeuvres complètes.* (2 vols., ed. B. Holmboe.) 1839, Christiana.

Works$_2$. *Oeuvres complètes.* (2 vols., ed. L. Sylow and S. Lie.) 1881, Christiana: reprinted 1965, New York; and Cleveland.

1826a. "Démonstration d'une expansion de laquelle la formule binôme est un cas particulier." *Journ. rei. ang. Math.*, 1(1826), 159–160; *Works*$_1$, 1, 31–32; *Works*$_2$, 1, 102–103.

1826b. "Untersuchungen über die Reihe $1 + (m/1)x + [m(m-1)/1.2]x^2 + \cdots$." *Journ. rei. ang. Math.*, 1(1826), 311–339; *Works*$_1$, 1, 66–92; *Works*$_2$, 1, 219–250; Ostwald's Klassiker, 71(ed. A. Wangerin. 1895, Leipzig).

1827a. "Aufgaben und Lehrsätze. 1 (von Herrn. N. H. Abel.)" *Journ. rei. ang. Math.*, 2(1827), 286; *Works*$_2$, 1, 618–619.

1828a. "Note sur un mémoire de M. L. Olivier, ..." *Journ. rei. ang. Math.*, 3(1828), 79–81; *Works*$_1$, 1, 111–113; *Works*$_2$, 1, 399–402.

1841a. "Mémoire sur une propriété générale d'une classe très-étendue de fonctions transcendantes." *Mém. prés. Acad. Roy. Sci. div. sav.*, (2)7(1841), 176–264; *Works*$_2$, 1, 145–211.

1902a. Niels Hendrik Abel. Mémorial publié à l'occasion du centenaire de sa naissance. 1902, Christiana.

Correspondence. "Correspondance d'Abel comprenant ses lettres et celles qui lui ont été adressées." *1902a,* 1–135 [separate pagination].

Letters. "Texte original des lettres écrites par Abel en Norwégien." *1902a,* 1–61 [separate pagination].

Series. "Sur les séries." Manuscript; *Works*$_2$, 2, 197–205.

d'Alembert, J. le R.

Pamphlets. Opuscules mathématiques. (8 vols.) 1761–80, Paris.

1747a. "Recherches sur la courbe que forme une corde tenduë mise en vibration." *Mém. Acad. Sci. Berlin*, 3(1747: publ. 1749), 214–219.

1747b. "Suite des recherches sur la courbe que forme une corde tenduë mise en vibration." *Mém. Acad. Sci. Berlin*, 3(1747: publ. 1749), 220–253.

1750a. "Addition au mémoire sur la courbe que forme une corde tenduë mise en vibration." *Mém. Acad. Sci. Berlin*, 6(1750: publ. 1752), 355–360.

1754a. Recherches sur différens points importans de systême du monde. (3 vols.) 1754–56, Paris.

1761a. "Recherches sur les vibrations des cordes sonores." *Pamphlets*, 1(1761). 1–73.

1767a. "Sur les principes métaphysiques du calcul infinitésimal." *Mélanges de litterature, d'histoire et de philosophie*$_2$, 5(1767 [and later printings], Amsterdam), 239–252.

1768a. "Nouvelles reflexions sur les vibrations des cordes sonores." *Pamphlets*, 4(1768), 128–155.

1768b. "Premier supplément au mémoire précédent." *Pamphlets*, 4(1768), 156–179.

1768c. "Réflexions sur les suites et sur les racines imaginaires." *Pamphlets*, 5(1768), 171–215.

Works. Oeuvres [*sic*]. (5 vols.) 1821–22, Paris.

Ampère, A. M.

1806a. "Recherches sur quelques points de la théorie des fonctions dérivées...." *Journ. Ec. Polyt.*, cah. 13, 6(1806), 148–181.

1826a. "Démonstration du théorème de Taylor...." *Ann. math. pur. appl.*, 17(1826–27), 317–330.

Arago, D. F. J.

Works. Oeuvres. (17 vols., ed. J. A. Barral.) 1854–62, Paris.

1838a. "Eloge historique de Joseph Fourier." *Mém. Acad. Roy. Sci.*, 14(1838), lxix–cxxxviii; *Works*, 1, 295–369. English trans. in *Biographies of Distinguished Scientific Men* (1857, London), 242–286.

1850a. "Poisson." 1850, manuscript; *Works*, 2, 593–698.

Arbogast, L. F. A.

1789a. "Essai sur de nouveaux principes du calcul différentiel et du calcul intégral, indépendantes de la théorie des infiniment petits et de celle des limites." 1789, manuscript. [MS 2089, library of the *Ecole Nationale des Ponts et Chaussées* in Paris.]

1791a. Mémoire sur la nature des fonctions arbitraires qui entrent dans les intégrales des équations aux différences partielles. 1791, St. Petersburg.

Bateman, H.

1907a. "The correspondence of Brook Taylor." *Bibl. math.*, (3)7(1906–07), 367–371.

Berkeley, G.

1734a. The Analyst. 1734, London; *Works*, 4(1951, London and Edinburgh), 53–102.

Bernoulli, D.

1753a. "Réflexions et éclaircissemens sur les nouvelles vibrations des cordes...." *Mém. Acad. Sci. Berlin*, 9(1753: publ. 1755), 147–172.

1753b. "Sur le mélange de plusieurs espèces de vibrations simples isochrones, ..." *Mém. Acad. Sci. Berlin*, 9(1753: publ. 1755), 173–195.

1758a. "Lettre de Monsieur Daniel Bernoulli, ..." *Journ. des sav.*, (1758), 157–166.

1772a. "De indole singulari serierum infinitarum...." *Novi Comm. Acad. Sci. Petrop.*, 17(1772: publ. 1773), 3–23.

1773a. "Theoria elementaria serierum, ..." *Novi Comm. Acad. Sci. Petrop.*, 18(1773: publ. 1774), 3–23.

Bernoulli, J.

Works. Opera Omnia. (4 vols.) 1742, Lausanne and Geneva: reprinted 1968, Hildesheim.

1727a. "Theoremata selecta pro conservatione virium vivarum demonstrandi...." *Comm. Acad. Sci. Petrop.*, 2(1727: publ. 1729), 200–207; *Works*, 3, 124–130.

1728a. "Meditationes de chordis vibrantibus." *Comm. Acad. Sci. Petrop.*, 3(1728: publ. 1732), 13–28; *Works*, 3, 198–210.

Bertrand, J. L. F.

1842a. "Règles sur la convergence des séries." *Journ. math. pur. appl.*, (1)7(1842), 35–54.

1870a. [Review of Valson *1868a.*] *Bull. sci. math.*, (1)1(1870), 105–117.

1904a. "Eloge de Augustin-Louis Cauchy." *Mém. Acad. Roy. Sci.*, 47(1904), clxxxiii–ccv.

Calculus. Traité de calcul différentiel et de calcul intégral. (2 vols.) 1864–70, Paris: reprinted Cleveland.

Bessel, F. W.

Papers. Gesammelte Abhandlungen. (3 vols., ed. R. Engelmann.) 1875–76, Leipzig.

1816a. "Analytische Auflösung der Kepler'schen Aufgabe." *Abh. Akad. Wiss. Berlin*, (1816–17: publ. 1819), math. Kl., 49–55; *Papers*, 1, 17–20.

1839a. "Über den Ausdruck einer Function ϕx durch Cosinusse und

Sinusse des Vielfachen von *x*." *Astron. Nachr.*, 16(1839), cols. 229–238; *Papers*, 2, 393–398.

See also Gauss *1880a*.

Biermann, K. R.

1959a. [Ed.] "Johann Peter Gustav Lejeune-Dirichlet. Dokumente für sein Leben und Wirken. (Zum 100. Todestag.)" *Abh. Dtsch. Akad. Wiss. Berlin, Kl. Math. Phys. Techn.*, (1959), no. 2.

1966a. "Karl Weierstrass. Ausgewählte Aspekte seiner Biographie." *Journ. rei. ang. Math.*, 223(1966), 191–220.

Biot, J. B.

1816a. *Traité de physique expérimentale et mathématique*. (4 vols.) 1816, Paris.

Bôcher, M.

1906a. "Introduction to the theory of Fourier's series." *Ann. of Math.*, (2)7(1906), 81–152.

Bockstaele, P. A.

1966a. "Nineteenth century discussions in Belgium on the foundations of the calculus." *Janus*, 53(1966), 1–16.

du Bois Reymond, P. D. G.

1870a. *Antrittsprogramm, enthaltend neue Lehrsätze über die Summen unendlicher Reihen*.... 1870, Berlin; Ostwald's Klassiker, 185(ed. P. E. B. Jourdain. 1912, Leipzig), 3–42.

1871a. "Notiz über einen Cauchy'schen Satz, die Stetigkeit von Summen unendlicher Reihen betreffend." *Math. Ann.*, 4(1871), 135–137.

1872a. "Eine neue Theorie der Convergenz und Divergenz von Reihen mit positiv Gliedern...." *Journ. rei. ang. Math.*, 76(1872), 61–91.

1876a. "Untersuchungen über die Convergenz und Divergenz der Fourier'schen Darstellungsformeln." *Abh. Akad. Wiss. Munich, math.-phys. Kl.*, 12(1876), part 2, i–xxiv and 1–102; Ostwald's Klassiker, 186(ed. P. E. B. Jourdain. 1913, Leipzig).

Bolzano, B. P. J. N.

1816a. *Der binomische Lehrsatz*.... 1816, Prague.

1817a. *Rein analytischer Beweis des Lehrsatzes*.... 1817, Prague; *Abh. Gesell. Wiss. Prague*, (3)5(1814–17: publ. 1818), 1–60 [separate pagination]; Ostwald's Klassiker, 153(ed. P. E. B. Jourdain. 1905, Leipzig), 3–43. French trans. in *Rev. hist. sci. appl.*, 17(1964), 136–164. [Also other reissues and translations, including in Kolman *1955a*.]

1817b. *Die drei Probleme der Rectification, der Complanation und die Cubirung*.... 1817, Leipzig; *Writings*, 5, 67–138 [with notes by J. Vojtech].

1836a. Lebensbeschreibung (Autobiographie) des Dr. B. Bolzano.... 1836, Sulzbach. [Edited for publication by M. J. Fesl.]

1851a. Paradoxien der Unendlichen. (Ed. F. Přihonský.) 1851, Leipzig. [Various reissues.]

1950a. Paradoxes of the Infinite. [English trans., with notes, of *1851a* by D. Steele.] 1950, London.

Numbers. Theorie der reellen Zahlen im Bolzanos handschriftlichen Nachlasse. (Ed. K. Rychlik.) 1962, Prague.

Writings. Schriften. (5 vols. [so far], ed. Königliche Böhmische Gesellschaft der Wissenschaften.) 1930–48, Prague.

Boncompagni, B.

1869a. [Review of Valson *1868a.*] *Bull. bibl. storia sci. mat. fis.*, 2(1869), 1–102.

Bonnet, P. O.

1843a. "Note sur la convergence et la divergence des séries." *Journ. math. pur. appl.*, (1)8(1843), 73–109.

1850a. "Sur la théorie générale des séries." *Mém. sav. étr., Acad. Roy. Belgique*, 23(1850), 1–116.

1879a. "Note sur la formule qui sert de fondement à une théorie des séries trigonométriques." *Bull. sci. math.*, (2)3, part 1 (1879), 480–484.

Borel, E. F. E. J.

1899a. "Mémoire sur les séries divergentes." *Ann. sci. Ec. Norm. Sup.*, (3)16(1899), 9–136.

1912a. [Ed.] "Recherches contemporaines sur la théorie des fonctions." *Enc. sci. math.*, article II 12 (1912, Paris).

Boyer, C. B.

1949a. The Concepts of the Calculus. 1949, New York: reprinted 1959, New York [as *The History of the Calculus and Its Conceptual Development*].

1956a. History of Analytic Geometry. 1956, New York.

Brill, A. and Nöther, M.

1893a. "Die Entwicklung der Theorie der algebraischen Funktionen in älterer und neuerer Zeit." *Jsbr. dtsch. Math.-Ver.*, 3(1892–93: publ. 1894), i–xxiii and 109–566.

Burkhardt, H. F. K. L.

1908a. "Entwicklungen nach oscillierenden Functionen und Integration der Differentialgleichungen der mathematischen Physik." *Jsbr. dtsch. Math.-Ver.*, 10, part 2 (1901–08).

1911a. "Über dem Gebrauch divergenter Reihen in der Zeit von 1750–1860." *Math. Ann.*, 70(1910–11), 169–206.

Cajori, F.

1893a. "Evolution of criteria of convergence." *Bull. New York Math. Soc.*, 2(1892–93: publ. 1893), 1–10.

1928a. "The early history of partial differential equations and of partial differentiation." *Amer. Math. Mon.*, 35(1928), 459–467.

1929a. A History of Mathematical Notations. Vol. 2: Notations Mainly in Higher Mathematics. 1929, Chicago.

Cantor, G. F. L. P.

Papers. Gesammelte Abhandlungen. Mathematischen und philosophischen Inhalts. (Ed. E. Zermelo.) 1932, Berlin: reprinted 1962, Hildesheim.

1882a. "Über unendliche, lineare Punktmannigfaltigkeiten." [Part 3.] *Math. Ann.*, 20(1882), 113–121; *Papers*, 149–157.

Carnot, L. N. M.

Calculus. Réflexions sur la métaphysique du calcul infinitésimal. First edition, 1797, Paris. Second edition, 1813, Paris. [Various reissues and translations.]

Carruccio, E.

1957a. "I fondamenti dell'analisi matematica nel pensiero di Agostino Cauchy." *Boll. Un. Mat. Ital.*, (3)12(1957), 298–307; *Rend. Sem. Mat. Univ. Polit. Torino*, 16(1956–57), 205–216.

Casorati, F.

1868a. Teoria delle funzioni di variabili complesse. 1868, Pavia.

Catalan, E.

1860a. Traité élémentaire des séries. 1860, Paris.

Cauchy, A. L.

Works. Oeuvres complètes. (12 + 14 [so far] vols., ed. Académie des Sciences.) 1882–   , Paris.

1814a. "Mémoire sur les intégrales définies." *Mém. prés. Acad. Roy. Sci. div. sav.*, (2)1(1827), 601–799; *Works*, (1)1, 319–506.

1815a. "Théorie de la propagation des ondes à la surface d'un fluide pesant d'une profondeur indéfinie." *Mém. prés. Acad. Roy. Sci. div. sav.*, (2)1(1827), 3–312; *Works*, (1)1, 4–318.

1817a. "Sur une loi de réciprocité qui existe entre certaines fonctions." *Bull. sci. Soc. Philom. Paris*, (1817), 121–124; *Works*, (2)2, 223–227.

1818a. "Second note sur les fonctions réciproques." *Bull. sci. Soc. Philom. Paris*, (1818), 178–181; *Works*, (2)2, 228–232.

1821a. Cours d'analyse de l'Ecole Royale Polytechnique. 1re partie: analyse algébrique. [No others published.] 1821, Paris; *Works*, (2)3. German transs.: 1828, Königsberg; 1895, Berlin. Russian trans.: 1864, Leipzig.

1822a. "Sur le développement des fonctions en série et sur l'intégration des équations différentielles, ou aux différences partielles." *Bull. sci. Soc. Philom. Paris*, (1822), 49–54; *Works*, (2)2, 276–282.

1822b. "Mémoire sur les intégrales définies...." *Bull. sci. Soc. Philom. Paris*, (1822), 161–174; *Works*, (2)2, 283–299.

1823a. Résumé des leçons données à l'Ecole Royale Polytechnique sur le calcul infinitésimal. Tome premier. [No others published.] 1823, Paris; *Works*, (2)4, 5-261. Russian trans.: 1831, St.Petersburg.

1823b. "Sur l'intégration des équations linéaires aux dérivées partielles et à coefficiens constans." *Journ. Ec. Polyt.*, cah. 19, 12(1823), 510–592; *Works*, (2)1, 275–357.

1825a. Mémoire sur les intégrales prises entre des limites imaginaires. 1825, Paris; *Bull. sci. math.*, (1)7(1874), 265–304, and (1)8(1875), 43–55 and 148–159. German trans. in Ostwald's Klassiker, 112(ed. P. Stäckel. 1900, Leipzig).

1826a. "Mémoire sur les développements des fonctions en séries périodiques." *Mém. Acad. Roy. Sci.*, 6(1823: publ. 1827), 603–612; *Works*, (1)2, 12–19.

1826–30a. Exercices de Mathématiques. (4 vols. and 1 installment.) 1826–30, Paris; *Works*, (2)6–9.

1826b. "Sur les formules de Taylor et Maclaurin." *1826–30a*, 1(1826), 25–28; *Works*, (2)6, 38–42.

1826c. "De l'influence que peut avoir, sur la valeur d'une intégrale double, l'ordre dans lequel on effectue les intégrations." *1826–30a*, 1(1826), 85–94; *Works*, (2)6, 113–123.

1826d. "Sur les divers ordres de quantités infiniment petites." *1826–30a*, 1(1826), 145–150; *Works*, (2)6, 184–190.

1827a. "De la différentiation sous le signe ∫." *1826–30a*, 2(1827), 125–139; *Works*, (2)7, 160–176.

1827b. "Sur la convergence des séries." *1826–30a*, 2(1827), 221–232; *Works*, (2)7, 267–279.

1827c. "Sur les résidues des fonctions exprimées par des intégrales définies." *1826–30a*, 2(1827), 341–376; *Works*, (2)7, 393–430.

1829a. Leçons sur le calcul différentiel. 1829, Paris; *Works*, (2)4, 263–609. German trans.: 1836, Brunswick.

1833a. Résumés analytiques. 1833, Turin; *Works*, (2)10, 5–184.

1833b. Quelques mots aux hommes de bon sens et de bonne foi. First edition, 1833, Montpellier. Second edition, 1833, Prague.

1835a. Nouveaux exercices des mathématiques. 1835, Prague; *Works*, (2)10, 185–464.

1840–47a. Exercices d'analyse et de physique mathématique. (4 vols.) 1840–47, Paris; *Works*, (2)11–14.

1841a. " Rapport sur un mémoire de M. Broch, relatif à une certaine classe d'intégrales." *Compt. Rend. Acad. Roy. Sci.*, 12 (1841), 847–850; *Works*, (1)6, 146–149.

1843a. " Mémoire sur l'analyse infinitésimal." *Compt. Rend. Acad. Roy. Sci.*, 17(1843), 275–279; *Works*, (1)8, 11–17.

1844a. " Mémoire sur l'analyse infinitésimal." *1840–47a*, 3(1844), 5–49; *Works*, (2)13, 9–58.

1844b. " Mémoire sur les fonctions continuées ou discontinuées." *Compt. Rend. Acad. Roy. Sci.*, 18(1844), 116–130; *Works*, (1)8, 145–160.

1844c. " Sur la convergence des séries multiples." *Compt. Rend. Acad. Roy. Sci.*, 19(1844), 1433–1436; *Works*, (1)8, 386–389.

1844d. " Mémoire sur diverses formules relatives à la théorie des intégrales définies...." *Journ. Ec. Polyt.*, cah. 28, 17(1844), 147–248; *Works*, (2)1, 467–567.

1853a. " Note sur les séries convergentes dont les divers termes sont des fonctions continuées d'une variable réelle ou imaginaire, entre des limites données." *Compt. Rend. Acad. Roy. Sci.*, 36(1853), 454–459; *Works*, (1)12, 30–36.

Césaro, E.

1890a. " Sur la multiplication des séries." *Bull. sci. math.*, (2) 4, part 1(1890), 114–120.

Champollion-Figeac, J. J.

1844a. Fourier et Napoléon, l'Egypte et les cents jours. 1844, Paris.

Clairaut, A. C.

1754a. " Mémoire sur l'orbite apparente du soleil autour de la terre, ..." *Mém. Acad. Roy. Sci.*, (1754: publ. 1759), 521–564.

Collingwood, E. F.

1959a. " Emile Borel." *Journ. Lond. Math. Soc.*, 34(1959), 488–512.

Cousin, V.

1831a. Notes biographiques pour faire suite à l'éloge de M. Fourier. 1831, Paris.

Crelle, A. L.

1813a. Versuch einer rein algebraischen.... Erster Band [no others published]. 1813, Göttingen.

1829a. " Démonstration nouvelle du théorème du binôme, ..." *Journ. rei. ang. Math.*, 4(1829), 305–308.

1829b. " Necrologie." *Journ. rei. ang. Math.*, 4(1829), 402–404.

1830a. " Mémoire sur la convergence de la série du binôme...." *Journ. rei. ang. Math.*, 5(1830), 187–196.

Darboux, J. G.

1875a. "Sur les fonctions discontinues." *Ann. sci. Ec. Norm. Sup.*, (2)4(1875), 57–112.

Dedekind, J. W. R.

1876a. "Bernhard Riemann's Lebenslauf." Riemann *Works*$_1$, 507–526; *Works*$_2$, 539–558.

Delambre, J. B. J.

1812a. "Notice sur la vie et l'ouvrage de M. Lagrange." *Hist. cl. sci. math. Inst. Fr.*, 13(1812: publ. 1814), xxxiv–lxxx; Lagrange *Works*, 1, ix–li.

Dickstein, S.

1899a. "Zur Geschichte der Prinzipien der Infinitesimalrechnung...." *Abh. Gesch. Math.*, 9(1899), 65–79.

Dirichlet, J. P. G. Lejeune-

Works. Gesammelte Werke. (2 vols., ed. L. Kronecker and L. Fuchs.) 1889–97, Berlin.

1829a. "Sur la convergence des séries trigonométriques qui servent à représenter une fonction arbitraire entre des limites données." *Journ. rei. ang. Math.*, 4(1829), 157–169; *Works*, 1, 117–132.

1830a. "Solution d'une question relative à la théorie mathématique de la chaleur." *Journ. rei. ang. Math.*, 5(1830), 287–295; *Works*, 1, 161–172.

1835a. "Über eine neue Anwendung bestimmter Integrale auf die Summation endlicher oder unendlicher Reihen." *Abh. Akad. Wiss. Berlin*, (1835), math.-phys. Kl., 391–407; *Works*, 1, 237–256.

1837a. "Über die Darstellung ganz willkürlicher Funktionen durch Sinus- and Cosinusreihen." *Report. Phys.*, 1(1837), 152–174; *Works*, 1, 133–160; Ostwald's Klassiker, 116(ed. H. Liebmann. 1900, Leipzig), 3–34.

1837b. "Sur les séries dont le terme général dépend de deux angles, ..." *Journ. rei. ang. Math.*, 17(1837), 35–56; *Works*, 1, 283–306.

1837c. "Sur l'usage des intégrales définies dans la sommation des séries finies ou infinies." *Journ. rei. ang. Math.*, 17(1837), 57–67; *Works*, 1, 257–270.

1837d. "Beweis des Satzes, ..." *Abh. Akad. Wiss. Berlin*, (1837), math.-phys. Kl., 45–81 [*sic*: actually 71]; *Works*, 1, 313–342.

1862a. "Démonstration d'un théorème d'Abel...." *Journ. math. pur. appl.*, (2)7(1862), 253–255; *Works*, 2, 303–306.

1904a. G. Lejeune-Dirichlet's Vorlesungen über die Lehre von den einfachen und mehrfachen bestimmten Integralen. (Ed. G. Arendt.) 1904, Brunswick.

Dirksen, E. H.

1827a. "Über die Darstellung beliebiger Funktionen mittelst Reihen, ..." *Abh. Akad. Wiss. Berlin*, (1827), math.-phys. Kl., 85–113.

1829a. "Über die Convergenz einer nach den Sinussen und Cosinussen der Vielfachen eines Winkels fortschreitenden Reihe." *Journ. rei. ang. Math.*, 4(1829), 170–178.

1829b. "Über die Summe einer, nach den Sinussen und Cosinussen der Vielfachen eines Winkels fortschreitenden, Reihe." *Abh. Akad. Wiss. Berlin*, (1829), math.-phys. Kl., 169–188.

1829c. [Review of first German trans. of Cauchy *1821a.*] *Jahrb. wissensch. Krit.*, (1829), cols. 211–222 [second pagination].

Dubbey, J. L.

1966a. "Cauchy's contribution to the establishment of the calculus." *Ann. of Sci.*, 22(1966), 61–67.

Duhamel, J. M. C.

1839a. "Nouvelle règle pour la convergence des séries." *Journ. math. pur. appl.*, (1)4(1839), 214–221.

Dunnington, G. W.

1955a. *Carl Friedrich Gauss: Titan of Science. A Study of His Life and Work.* 1955, New York.

Encyclopaedia

Encyclopaedia. Encyclopédie ou dictionnaire raisonné des sciences, des arts et des metiers. (28 vols., ed. D. Diderot and J. le R. d'Alembert.) 1751–65, Paris, Neuchâtel and Amsterdam; 1772, Geneva.

Eneström, G.

1894a. "Note upon the rules of convergence in the eighteenth century." *Bull. New York Math. Soc.*, 3(1893–94: publ. 1894), 186–187.

1905a. "Über eine von Euler aufgestellte allgemeine Konvergenzbedingung." *Bibl. math.*, (3)6(1905), 186–189.

Euler, L.

Works. Opera Omnia. (3 series [all as yet incomplete], ed. Societas Scientarum Naturalium Helveticae.) 1911–   , Leipzig, Berlin and Zurich.

1729a. "De serierum determinatione seu nova methodus inveniendi generalis serierum." *Novi Comm. Acad. Sci. Petrop.*, 3(1750–51: publ. 1753), 36–85; *Works*, (1)14, 463–515.

1734a. "De progressionibus harmonicis observationes." *Comm. Acad. Sci. Petrop.*, 7(1734–35: publ. 1740), 150–161; *Works*, (1)14, 87–100.

1734b. "De infinitis curvis eiusdem generis, seu methodus inveniendi aequationes pro infinitis curvis eiusdem generis." *Comm. Acad. Sci. Petrop.*, 7(1734–35: publ. 1740), 174–189 and 180–183 [*sic*: incorrect pagination]; *Works*, (1)22, 36–56.

1734c. "Additamentum ad dissertationem de infinitis curvis eiusdem generis." *Comm. Acad. Sci. Petrop.*, 7(1734–35: publ. 1740), 184–200; *Works*, (1)22, 57–75.

1736a. "Methodus universalis serierum convergentium summas quam proxime inveniendi." *Comm. Acad. Sci. Petrop.*, 8(1736: publ. 1741), 3–9; *Works*, (1)14, 101–107.

1747a. "De propagatione pulsuum per medium elasticum." *Novi Comm. Acad. Sci. Petrop.*, 1(1747–48: publ. 1750), 67–105; *Works*, (2)10, 98–131.

1748a. "Sur la vibration des cordes." [French trans. of *1749b.*] *Mém. Acad. Sci. Berlin*, 4(1748: publ. 1750), 69–85; *Works*, (2)10, 63–77.

1748b. Introductio ad analysin infinitorum. (2 vols.) 1748, Lausanne; *Works*, (1)8–9. [Many reissues and translations.]

1749a. Recherches sur la question des inégalités du mouvement de Saturne et de Jupiter. 1749, Paris; *Works*, (2)25, 45–157.

1749b. "De vibratione chordarum exercitato." *Nova Acta Erud.*, (1749), 512–527; *Works*, (2)10, 50–62.

1753a. "Remarques sur les mémoires précédens de M. Bernoulli." *Mém. Acad. Sci. Berlin*, 9(1753: publ. 1755), 196–222; *Works*, (2)10, 233–254.

1754a. "Subsidium calculi sinuum." *Novi Comm. Acad. Sci. Petrop.*, 5(1754–55: publ. 1760), 164–204; *Works*, (1)14, 542–584.

1754b. "De seriebus divergentibus." *Novi Comm. Acad. Sci. Petrop.*, 5(1754–55: publ. 1760), 205–237; *Works*, (1)14, 585–617.

1755a. Institutiones calculi differentialis. 1755, St. Petersburg; *Works*, (1)10. [Several reissues.]

1763a. "De usu functionum discontinuarum in analysi." *Novi Comm. Acad. Sci. Petrop.*, 11(1763: publ. 1768), 67–102; *Works*, (1)23, 74–91.

1765a. "Sur le mouvement d'une corde, qui au commencement n'a été ébranlée que dans une partie." *Mém. Acad. Sci. Berlin*, (1765: publ. 1767), 307–334; *Works*, (2)10, 426–450.

1765b. "Eclaircissemens sur le mouvement des cordes vibrantes." *Miscell. Taurin.*, 3(1762–65: publ. 1766), cl. math., 1–26; *Works*, (2)10, 377–396.

1769a. "De formulis integralibus duplicatis." *Novi Comm. Acad. Sci. Petrop.*, 14(1769: publ. 1770), 72–103; *Works*, (1)17, 289–315.

1772a. "De chordis vibrantibus disquisitio ulterior." *Novi Comm. Acad. Sci. Petrop.*, 17(1772: publ. 1773), 381–409; *Works*, (2)11, sect. 1, 62–80.

1773a. "Summatio progressionum
$\sin\phi^\lambda + \sin 2\phi^\lambda + \sin 3\phi^\lambda + \cdots + \sin n\phi^\lambda$
$\cos\phi^\lambda + \cos 2\phi^\lambda + \cos 3\phi^\lambda + \cdots + \cos n\phi^\lambda$."
Novi Comm. Acad. Sci. Petrop., 18(1773: publ. 1774), 24–36; *Works*, (1)15, 168–184.

1777a. "Disquisitio ulterior super seriebus secundum multipla cuiusdem anguli progredientibus." *Nova Acta Acad. Sci. Petrop.*, 11(1793: publ. 1798), 114–132; *Works*, (1)16, sect. 1, 333–355.

Integration. Institutiones calculi integralis. (3 vols.) 1768–70, St. Petersburg; *Works*, (1)11–13. [Several reissues, including a 4th volume.]

Series. "Series maxime idoneae pro circuli quadratura proxime invenienda." Manuscript; *Opera Postuma*.... (2 vols., ed. P. H. and N. Fuss. 1862, St. Petersburg), 1, 288–298; *Works*, (1)16, sect. 2, 267–283.

Fleckenstein, J. O.

1946a. "Die Taylor'schen Formel bei Johann I Bernoulli." *Elem. der Math.*, 1(1946), 13–17.

Folta, J.

1961a. "N. I. Lobačevsky a B. Bolzano." *Pok. mat., fys. ast.*, 6(1961), 283–284.

Fourier, J. B. J.

Works. Oeuvres. (2 vols., ed. G. Darboux.) 1888–90, Paris.

1807a. "Sur la propagation de la chaleur." 1807, manuscript. [MS 1851, library of the *Ecole Nationale des Ponts et Chaussées*, Paris. To appear in the forthcoming Grattan-Guinness and Ravetz *Fourier*.]

1808a. "Note sur la convergence de la série $\sin x - \frac{1}{2}\sin 2x + \frac{1}{3}\sin 3x - \frac{1}{4}\sin 4x + \cdots$." 1808, manuscript. [With *1807a*.]

1811a. "Théorie du mouvement de la chaleur dans les corps solides." *Mém. Acad. Roy. Sci.*, 4(1819–20: publ. 1824), 185–555; and 5(1821–22: publ. 1826), 153–246. Second part: *Works*, 2, 3–94.

1816a. "Théorie de la chaleur." *Ann. chim. et phys.*, 3(1816), 350–375.

1818a. "Question d'analyse algébrique." *Bull. sci. Soc. Philom. Paris*, (1818), 61–67; *Works*, 2, 241–253.

1822a. La théorie analytique de la chaleur. 1822, Paris: reprinted 1883, Breslau; *Works*, 1. English trans.: 1878, Cambridge: reprinted 1955, New York. German trans.: 1884, Berlin.

1829a. "Sur la théorie analytique de la chaleur." *Mém. Acad. Roy. Sci.*, 8(1829), 581–622; *Works*, 2, 145–181.

Fréchet, R. M.

1940a. "Biographie du mathématicien alsacien Arbogast." *Thalés*, 4(1940), 43–55.

Funk, P.

1967a. [With W. Frank.] "Bolzano als Mathematiker." *Sitz.-Ber. Öst. Akad. Wiss. Wien, phil.-hist. Kl.*, 252(1967), part 5, 121–134.

Fuss, P. H.

1843a. [Ed.] *Correspondance mathématique et physique de quelques célèbres géomètres du XVIII siècle.* (2 vols.) 1843, St. Petersburg.

Gagaev, B. M. (Гагаев, Б. М.)

1952a. "Обобщение Н. И. Лобачевским Интеграле Фуре." *Сто двадцать пять лет неевклидовой геометрии Лобачевского 1826–1951* (ed. A. P. Norden. 1952, Moscow and Leningrad), 79–86.

Galois, E.

1830a. "Notes sur quelques points d'analyse." *Ann. math. pur. appl.*, 21(1830), 182–184; *Journ. math. pur. appl.*, (1)11(1846), 392–394; *Oeuvres mathématiques* (ed. E. Picard. 1897, Paris), 9–10.

Gauss, K. F.

Works. Werke. (12 vols., ed. Königliche Preussische Gesellschaft der Wissenschaften.) 1863–1933, Leipzig: reprinted 1969, Hildesheim.

1797a. "Neue Methode die Summe der divergierenden Reihe $1 - 1 + 2 - 6 + 24 -$ etc. [=] 0,5963 ... zu finden." 1797, manuscript; *Works*, 10, part 1, 382–389.

1799a. Demonstratio nova theorematis omnem functionem algebraicam.... 1799, Helmstedt; *Works*, 3, 1–56. [Arts. 1–12 in Euler *Works*, (1)6, 151–169.]

1809a. ["Darstellung von diskontinuerlichen Funktionen."] 1809, manuscript; *Works*, 10, part 1, 398–399.

1813a. "Disquisitio generales circa seriem infinitam $1 + \frac{\alpha.\beta}{1.\gamma}x + \frac{\alpha(\alpha+1).\beta(\beta+1)}{1.2.\gamma(\gamma+1)}x^2 + \frac{\alpha(\alpha+1)(\alpha+2).\beta(\beta+1)(\beta+2)}{1.2.3.\gamma(\gamma+1)(\gamma+2)}x^3 + \cdots$. Pars 1." [No others published.] *Comm. Soc. Reg. Sci. Göttingen Rec.*, 2(1811–13: publ. 1813), cl. math., 1–46 [separate pagination]; *Works*, 3, 123–162. German trans.: 1888, Berlin.

1815a. "Demonstratio nova altera theorematis omnem functionem algebraicam...." *Comm. Soc. Reg. Sci. Göttingen Rec.*, 3(1814–15: publ. 1816), cl. math., 107–134; *Works*, 3, 31–56.

1816a. "Theorematis de resolubitate...." *Comm. Soc. Reg. Sci. Göttingen Rec.*, 3(1814–15: publ. 1816), cl. math., 135–142; *Works*, 3, 57–64. [German trans. of *1799a*, *1815a* and *1816a* in Ostwald's Klassiker, 14(ed. E. Netto. 1904, Leipzig), 1–67.]

1880a. Briefwechsel zwischen Gauss und Bessel. (Ed. Königliche Preussische Akademie der Wissenschaften.) 1880, Leipzig.

Equation. "Determinatio serie nostrae per aequationem differentialem secundi ordinis." Manuscript; *Works*, 3, 207–230.

Series. "Grundbegriffe der Lehre von der Reihen." Manuscript; *Works*, 10, part 1, 390–395.

Gibson, G. A.

1893a. "On the history of the Fourier series." *Proc. Edin. Math. Soc.*, 11(1892–93), 137–166.

Gloden, A.

1950a. "Les développements de la théorie des séries depuis le debut du

19e siècle jusqu'à nos jours." *Arch. Inst. Grand-Ducal Luxembourg, sect. sci. nat. phys. math.*, (2)19(1950), 205–220.

Grattan-Guinness, I.

1969a. "Joseph Fourier and the revolution in mathematical physics." *Journ. Inst. Maths. Applics.*, 5(1969), 230–253.

1970a."An unpublished paper by Georg Cantor: Principien einer Theorie der Ordnungstypen. Erste Mittheilung." *Acta Math.*, 124(1970), 65–107.

1970b."Berkeley's criticism of the calculus as a study in the theory of limits." *Janus*, 56(1970).

1970c. "Bolzano, Cauchy and the 'new analysis' of the early nineteenth century." *Arch. hist. exact sci.*, 6(1970), 372–400.

Grattan-Guinness, I. and Ravetz, J. R.

Fourier. Joseph Fourier 1768–1830. A Survey of His Life and Work, Based on a Critical Edition of his Monograph on the Propagation of Heat, Presented to the Institut de France in 1807. Probably 1971, Cambridge, Mass.

Grimsley, R.

1963a. Jean d'Alembert (*1717–1783*). 1963, London.

Hankel, H.

1869a. Die Entdeckung der Mathematik in den letzten Jahrhunderten. Ein Vortrag. 1869, Tübingen.

1870a. Untersuchungen über die unendlich oft oscillierenden und unstetigen Functionen. 1870, Tübingen; *Math. Ann.*, 20(1882), 63–112; Ostwald's Klassiker, 153(ed. P. E. B. Jourdain. 1905, Leipzig), 44–102.

1871a. "Grenze." *Allg. Enc. Wiss. Künste*, sect. 1, part 90 (1871, Leipzig), 185–211.

Hardy, G. H.

1918a. "Sir George Stokes and the concept of uniform convergence." *Proc. Camb. Phil. Soc.*, 19(1918), 148–156.

1949a. Divergent Series. 1949, Oxford.

Harnack, A. C. G.

1888a. "Über Cauchy's zweiten Beweis für die Convergenz der Fourier'schen Reihe und eine damit verwandte ältere Methode von Poisson." *Math. Ann.*, 32(1888), 175–202.

Heine, E. H.

1870a. "Über trigonometrische Reihen." *Journ. rei. ang. Math.*, 71(1870). 353–365.

Hofmann, J. E.

1959a. "Eulers erste Reihestudien." *Sammelband der zu Ehren des 250. Geburtstages Leonhard Eulers* (1959, Berlin), 139–207.

l'Huilier, S. A. J.

1786a. Exposition élémentaire des principes des calculs supérieures. 1786, Berlin.

Jarník, V.

1922a. "O funkci Bolzanové." *Čas. pěst. mat. fys.*, 51(1922), 248–264.

1931a. "Bolzanova Functionenlehre." *Čas. pěst. mat. fys.*, 60(1931), 240–265.

1953a. "Bernard Bolzano o základy matematické analysy." *Zdeňku Nejedlému Československa Akademie Věd* (1953, Prague), 450–458.

1961a. "Bernard Bolzano (October 5, 1781–December 18, 1848)." *Czech. Math. Journ.*, 11(1961), 485–489.

Jordan, M. E. C.

Course. Cours d'Analyse de l'Ecole Polytechnique. (3 vols.) First edition, 1882–87, Paris. Second edition, 1892–96, Paris: reprinted Cleveland. Third edition, 1909–15, Paris.

Jourdain, P. E. B.

1905a. "The theory of functions with Cauchy and Gauss." *Bibl. math.*, (3)6(1905), 190–207.

1906a. "The development of the theory of transfinite numbers. Part 1. The growth of the theory of functions up to the year 1870." *Arch. Math. Phys.*, (3)10(1906), 254–281.

1909a. "The development of the theory of transfinite numbers. Part 2. Weierstrass (1840–1880)." *Arch. Math. Phys.*, (3)14(1908–09), 287–311.

1913a. "The origins of Cauchy's conceptions of a definite integral and of the continuity of a function." *Isis*, 1(1913), 661–703.

1913b. "The ideas of the 'fonctions analytiques' in Lagrange's early work." *Proc. 5th Int. Cong. Math.*, 2(1913, Cambridge), 540–541.

1917a. "The influence of Fourier's theory of the conduction of heat on the development of pure mathematics." *Scientia*, 22(1917), 245–254.

Klein, C. F.

1926a. Vorlesungen über die Entwicklung der Mathematik im 19. Jahrhundert. I Band. (Ed. R. Courant and O. Neugebauer.) 1926, Berlin: reprinted 1950, New York.

Klügel, G. S. [and others].

Dictionary. Mathematisches Wörterbuch.... Erste Abtheilung. Die reine

Mathematik. (5 parts and 2 supplements.) 1803–1836, Leipzig: reprinted 1968, Hildesheim.

1823a. "Summierbare Reihe." *Dictionary*, part 4(1823), 560–595.

1833a. "Binomischer Lehratz." *Dictionary*, suppl. 1(1833), 283–341.

1833b. "Convergenz der Reihen." *Dictionary*, suppl. 1(1833), 416–456.

Knopp, K.

Series. Theory and Application of Infinite Series. [English trans. by Miss R. C. Young.] First edition, 1928, London and Glasgow. Second edition, 1951, London and Glasgow.

Kolman, E. (Кольман, Э.)

1955a. *Бернард Больцано*. 1955, Moscow. Czech. trans.: 1958, Prague. German trans.: 1963, Berlin.

Kowalewski, G.

1923a. "Über Bolzanos nichtdifferenzierbare stetige Funktion." *Acta Math.*, 40 (1923), 315–319.

Kummer, E. E.

1835a. "Über die Konvergenz und Divergenz der unendlichen Reihen." *Journ. rei. ang. Math.*, 13(1835), 171–184.

1836a. "Über die hypergeometrische Reihe

$$1+\frac{\alpha.\beta}{1.\gamma}x+\frac{\alpha(\alpha+1).\beta(\beta+1)}{1.2.\gamma(\gamma+1)}x^2+\frac{\alpha(\alpha+1)(\alpha+2).\beta(\beta+1)(\beta+2)}{1.2.3.\gamma(\gamma+1)(\gamma+2)}x^3+\cdots."$$

Journ. rei. ang. Math., 15(1836), 39–83 and 127–172.

1860a. "Gedächtnissrede auf Gustav Peter Lejeune-Dirichlet." *Abh. Akad. Wiss. Berlin*, (1860), 1–36; Dirichlet *Works*, 2, 309–344.

Lacroix, S. F.

Calculus. Traité élémentaire du calcul différentiel et du calcul intégral. First edition, 1802, Paris. Second edition, 1806, Paris. [Many later editions.]

Treatise. Traité du calcul différentiel et du calcul intégral. (3 vols.) First edition, 1797–1800, Paris. Second edition, 1810–19, Paris: reprinted Cleveland.

Lagrange, J. L.

Works. Oeuvres. (14 vols., ed. J. A. Serret, G. Darboux and others.) 1867–92, Paris: reprinted 1968, Hildesheim.

1754a. *Lettera di Luigi de la Grange Tournier*, ... 1754, Turin; *Works*, 7, 581–588.

1759a. "Recherches sur la nature, et la propagation du son." *Miscell. Taurin.*, 1(1759), cl. math., i–x and 1–112; *Works*, 1, 39–148.

1762a. ["Note sur la métaphysique du calcul infinitésimal."] *Miscell. Taurin.*, 2(1760–62: publ. 1762), cl. phil., 17–18 (footnote); *Works*, 7, 595–599.

1765a. "Solutions des différens problèmes du calcul intégral." *Miscell. Taurin.*, 3(1762–65: publ. 1766), cl. math., 179–380; *Works*, 1, 469–668.

1768a. "Nouvelle methode pour résoudre les équations littérales par le moyen des séries." *Mém. Acad. Sci. Berlin*, (1768: publ. 1770), cl. math., 251–326; *Works*, 3, 5–73.

1772a. "Sur une nouvelle espèce de calcul relatif à la différentiation et à l'intégration des quantités variables." *Nouv. Mém. Acad. Sci. Berlin*, (1772: publ. 1774), cl. math., 185–221; *Works*, 3, 439–470.

1772b. "Sur l'intégration des équations à différences partielles du premier ordre." *Nouv. Mém. Acad. Sci. Berlin*, (1772: publ. 1774), cl. math., 353–386; *Works*, 3, 547–575.

1774a. "Sur les intégrales particulières des équations différentielles." *Nouv. Mém. Acad. Sci. Berlin*, (1774: publ. 1776), cl. math., 197–276; *Works*, 4, 3–108.

1799a. "Discours sur l'objet de la théorie des fonctions analytiques." *Journ. Ec. Polyt.*, cah. 6, 2(1799), 232–235; *Works*, 7, 323–328.

1823a. J. L. Lagrange's mathematische Werke, 1–2(1823, Berlin). [German translation of *Functions*$_2$ and *Lessons* by A. L. Crelle.]

Functions$_1$. *Théorie des fonctions analytiques....* 1797, Paris; *Journ. Ec. Polyt.*, cah. 9, 3(1801), 1–277.

Functions$_2$. *Théorie des fonctions analytiques....* 1813, Paris; *Journ. Ec. Polyt.*, cah. 9, 3(1813) [*sic*], 1–383; *Works*, 9.

Lessons. "Leçons sur les calculs des fonctions." Lessons 1–20: *Séances Ec. Norm.*$_2$, 10(1801), 1–534; *Journ. Ec. Polyt.*, cah. 12, 5(1804), 1–324. Lessons 21–22: *Journ. Ec. Polyt.*, cah. 14, 7(suppl.)(1808), 1–90. Lessons 1–22: 1806, Paris; *Works*, 10.

Mechanics. Mécanique Analitique. First edition, 1788, Paris. Second edition, 2 vols., 1811–15, Paris; *Works*, 11–12.

Laplace, P. S.

Works. Oeuvres. (14 vols., ed. Académie des Sciences.) Vols. 1–7: 1843–47, Paris. Vols. 1–14: 1878–1912, Paris: reprinted 1969, Hildesheim.

1773a. "Recherches sur le calcul intégral aux différences partielles." *Mém. Acad. Roy. Sci.*, (1773: publ. 1777), 341–402; *Works*, 9, 3–68.

1779a. "Mémoire sur les suites." *Mém. Acad. Roy. Sci.*, (1779: publ. 1782), 207–309; *Works*, 10, 1–89.

1809a. "Sur les mouvements de la lumière dans les milieux diaphanes." *Mém. cl. sci. math. phys. Inst. Fr.*, 10(1809), 300–342; *Works*, 12, 265–298.

1809b. "Mémoire sur divers points d'analyse." *Journ. Ec. Polyt.*, cah. 15, 8(1809), 229–265; *Works*, 14, 178–214.

Essay. Essai philosphique sur la probabilité. First edition, 1814, Paris. [Several later editions.]

Mechanics. Traité de mécanique celeste. (5 vols.) 1799–1825, Paris; *Works*, 1–5.

Probability. Théorie analytique des probabilités. First edition, 1812, Paris. Second edition, 1814, Paris. Third edition, 1820, Paris: reprinted Cleveland; *Works*, 7.

Laptiev, B. L. (Лаптев, Б. Л.)

1959a. "О библиотечных записях книг и журналов, выданных Н. И. Лобачевскому." *Чсп. Мат. Наук*, 14(1959), part 5, 153–155.

Laugwitz, D.

1965a. "Bemerkungen zu Bolzanos Grössenlehre." *Arch. hist. exact sci.*, 2(1962–66), 398–409.

Laurent, H.

1862a. Théorie des séries. 1862, Paris.

Legendre, A. M.

1787a. "Mémoire sur l'intégration de quelques équations aux différences partielles." *Mém. Acad. Roy. Sci.*, (1787: publ. 1789), 309–351.

Exercises. Exercices du calcul intégral sur divers ordres de nombres transcendantes et sur les quadratures. (3 vols.) 1811–17, Paris.

Lipschitz, R. O. S.

1864a. "De explicatione per series trigonometricas . . . ." *Journ. rei. ang. Math.*, 63(1864), 296–308. French trans. in *Acta Math.*, 36(1913), 281–295.

Lobachevski, N. I. (Лобачевский, Н. И.)

Works. Полное собрание сочинений. (5 vols., ed. B. F. Kazan, A. L. Kolmogorov, and others.) 1946–51, Moscow and Leningrad.

1841a. "Über die Convergenz der unendlichen Reihen." *Meteorologische Beobachtungen . . . der Kaiserlichen Russischen Universität Kazan.* Heft 1, 1835–1836 [no others published], (1841, Kazan), 1–48; *Works*, 5, 163–218 [in a Russian trans., and with editorial notes].

Lunts, G. L. (Лунц, Г. Л.)

1949a. "О работах Н. И. Лобачевского по математическому анализу." *Ист.-мат. Исслед.*, (1)2(1949), 9–71.

1950a. "Аналитические Н. И. Лобачевского." *Чсп. мат. наук.*, 5(1950), part 1, 187–193.

Maclaurin, C.

1742a. Treatise on Fluxions. (2 vols.) 1742, Edinburgh.

Manheim, J. H.

1964a. The Genesis of Point Set Topology. 1964, New York.

Medvedev, F. A. (Медведев, Ф. А.)

1965a. Развитие теории множесть в XIX веке. 1965, Moscow.

Mertens, F.

1875a. "Über die Multiplicationsregel für zwei unendliche Reihen." *Journ. rei. ang. Math.*, 79(1875), 182–184.

Meyer, G. F.

1871a. Vorlesungen über die Theorie der bestimmten Integrale zwischen reellen Grenzen. 1871, Leipzig.

Mittag-Leffler, M. G.

1902a. "Une page de la vie de Weierstrass." *Compt. Rend. 2e Cong. Int. Math.* (1902, Paris), 131–153.

1912a. "Zur Biographie von Weierstrass." *Acta Math.*, 35(1912), 29–65.

1923a. "Die ersten 40 Jahre des Lebens von Weierstrass." *Acta Math.*, 39(1923), 1–57.

1927a. "Auszug aus einem Briefe von G. Mittag-Leffler an den Herausgeber dieser Zeitschrift." *Journ. rei. ang. Math.*, 157(1927), 12–14.

Moigno, F. N. M.

Calculus. Leçons de calcul différentiel et de calcul intégral. (4 vols.) 1840–61, Paris.

Monge, G.

1809a. "Construction de l'équation des cordes vibrantes." *Journ. Ec. Polyt.*, cah. 15, 8(1809), 118–145.

de Morgan, A.

1842a. The Differential and Integral Calculus. 1842, London.

1844a. "On divergent series, and various points of analysis connected with them." *Trans. Camb. Phil. Soc.*, 8(1842–47), 182–203.

1864a. "A theorem relating to neutral series." *Trans. Camb. Phil. Soc.*, 11(1866–69), 190–202.

Neumann, C. G.

1914a. "Über die Dirichlet'sche Theorie der Fourier'schen Reihen. . . ." *Abh. Gesell. Wiss. Leipzig*, 33(1914), math.-phys. Kl., 113–194.

Ohm, M.

Mathematics. Versuch eines vollständig consequenten Systems der Mathematik. (9 vols.) 1828–52, Berlin.

Olivier, L.

1827a. "Remarques sur les séries infinies et leur convergence." *Journ. rei. ang. Math.*, 2(1827), 31–44.

1828a. "Remarque de M. L. Olivier." *Journ. rei. ang. Math.*, 3(1828), 82.

Ore, O.

1957a. Niels Hendrik Abel—Mathematician Extraordinary. 1957, Minneapolis.

Osgood, W. F.

1901a. "Grundlagen der allgemeinen Theorie der analytischer Functionen einer complexen Grösse." *Enc. math. Wiss.*, article IIB (1901, Leipzig), 1–114.

Paplauskas, A. B. (Паплаускас, А. Б.)

1966a. Тригонометрические ряды от Эйлера до Лебега. 1966, Moscow.

von Paucker, M. G.

1851a. "Note relative à quelques règles sur la convergence des séries." *Journ. rei. ang. Math.*, 42(1851), 138–150.

Peacock, G.

1834a. "Report on the recent progress and present state of certain branches of analysis." *Report Brit. Ass. Adv. Sci.*, (1834), 185–352.

Pesin, I. N. (Песин, И. Н.)

1966a. Развитие понятия интеграла. 1966, Moscow.

Pincherle, S.

1880a. "Saggio di una introduzione alla teoria delle funzioni analitiche secondo i principii del Prof. C. Weierstrass." *Giorn. mat.*, (1)18(1880), 178–254 and 317–357.

Poisson, S. D.

1807a. "Mémoire sur la théorie du son." *Journ. Ec. Polyt.*, cah. 14, 7(1807), 319–392.

1808a. "Mémoire sur la propagation de la chaleur dans les corps solides." *Nouv. bull. sci. Soc. Philom. Paris*, 1(1808), 112–116; Fourier *Works*, 2, 213–221.

1814a. "Mémoire sur les intégrales définies." *Bull. sci. Soc. Philom. Paris*, (1814), 185–188; Cauchy *Works*, (2)2, 194–198.

1820a. "Mémoire sur la manière d'exprimer les fonctions. . . ." *Journ. Ec. Polyt.*, cah. 16, 11(1820), 417–489.

1823a. "Mémoire sur la distribution de la chaleur dans les corps solides." *Journ. Ec. Polyt.*, cah. 19, 12(1823), 1–144.

1823b. "Addition au mémoire précédent...." *Journ. Ec. Polyt.*, cah. 19, 12(1823), 145–162.

1823c. "Suite du mémoire sur les intégrales définies...." *Journ. Ec. Polyt.*, cah. 19, 12(1823), 404–509.

1823d. "Mémoire sur le calcul numérique des intégrales définies." *Mém. Acad. Roy. Sci.*, 6(1823: publ. 1827), 571–602.

1835a. *Théorie mathématique de la chaleur*. 1835, Paris.

Pringsheim, A.

1890a. "Allgemeine Theorie der Convergenz und Divergenz von Reihen mit positiven Gliedern." *Math. Ann.*, 35(1890), 297–394.

1900a. "Zur Geschichte des Taylorschen Lehrsatzes." *Bibl. math.*, (3)1(1900), 433–479.

1905a. "Über ein Eulersches Konvergenzkriterium." *Bibl. math.*, (3)6(1905), 252–256.

Raabe, J. L.

1832a. "Untersuchungen über die Konvergenz und Divergenz der Reihen." *Zeitschr. Phys. Math.*, 10(1832), 41–74.

1834a. "Note zur Theorie der Convergenz und Divergenz der Reihen." *Journ. rei. ang. Math.*, 11(1834), 309–310.

1841a. "Note sur la théorie de la convergence et de la divergence des séries." *Journ. math. pur. appl.*, (1)6(1841), 85–88.

Ravetz, J. R.

1961a. "Vibrating strings and arbitrary functions." *Logic of Personal Knowledge: Essays Presented to M. Polanyi on His 70th Birthday* (1961, London), 71–88.

Reiff, R.

1889a. *Geschichte der unendlichen Reihen*. 1889, Tübingen.

Riemann, G. F. B.

Works. *Gesammelte mathematische Werke*. (Ed. H. Weber and R. Dedekind.) First edition, 1876, Leipzig: reprinted Cleveland. Second edition, 1892, Leipzig.

Supplement. *Nachträge*. (Ed. M. Nöther and W. Wirtinger.) 1902, Leipzig. [*Works*$_2$ and *Supplement* reprinted together: 1953, New York.]

Works (French). *Oeuvres mathématiques de Riemann*. [Partial trans. of *Works*$_2$.] (Ed. L. Laugel.) 1898, Paris: reprinted 1968, Paris; and Cleveland.

1851a. *Grundlagen der allgemeine Theorie der Functionen einer veränderlichen complexen Grösse*. 1851, Göttingen; 1867, Göttingen; *Works*$_1$, 3–47; *Works*$_2$, 3–48; *Works* (*French*), 1–60.

1866a. "Über die Darstellbarkeit einer Function durch eine trigonometrische Reihe." *Abh. Gesell. Wiss. Göttingen*, 13(1866–67: publ. 1868), math. Kl., 87–132; *Works*$_1$, 213–253; *Works*$_2$, 227–271; *Works* (*French*), 225–279.

1866b. "Über die Hypothesen, welche der Geometrie zu Grunde liegen." *Abh. Gesell. Wiss. Göttingen*, 13(1866–67: publ. 1868), math. Kl., 133–152; *Works*$_1$, 254–269; *Works*$_2$, 272–287; *Works* (*French*), 280–299.

Lectures. Vorlesungen über partielle Differentialgleichungen und deren Anwendung auf physicalische Fragen. (Ed. K. Hattendorff.) 1869, Brunswick.

van Rootselaar, B.

1964a. "Bolzano's theory of real numbers." *Arch. hist. exact sci.*, 2(1962–66), 168–180.

1969a. "Bolzano's corrections to his 'Functionenlehre'." *Janus*, 56(1969), 1–21.

Rychlik, K.

1961a. "La théorie des nombres réels dans un ouvrage posthume manuscrit de Bernard Bolzano." *Rev. hist. sci. appl.*, 14(1961), 313–327.

1964a. "Niels Hendrik Abel a Čechy." *Pok. mat., fys. ast.*, 9(1964), 317–319.

Sachse, A.

1880a. "Versuch eine Geschichte der Darstellung willkürlicher Funktionen einer Variabele durch trigonometrische Reihen." *Abh. Gesch. Math.*, 3(1880), 229–276. French trans. in *Bull. sci. math.*, (2)4, part 1(1880), 43–64 and 83–112.

Schlesinger, L.

Gauss. "Über Gauss' Arbeiten zur Funktionentheorie." First edition: *Materialien für eine wissenschaftliche Biographie von Gauss*, 3(1912, Leipzig). Second edition: Gauss *Works*, 10, part 2, sect. 2(1933).

Schönflies, A. M.

Sets. "Die Entwicklung der Lehre von den Punktmannigfaltigkeiten." Part $1,_1$: *Jsbr. dtsch. Math.-Ver.*, 8(1900), part 2. Part 2: *Jsbr. dtsch. Math.-Ver.*, Ergsbd. 2(1907); 1908, Leipzig. Part $1,_2$ [with H. Hahn]: *Entwicklung der Mengenlehre und ihrer Anwendungen.* 1913, Leipzig and Berlin: reprinted Cleveland.

Schönflies, A. M. and Baire, R.

1909a. "Théorie des ensembles." *Enc. sci. math.*, article I7(1909, Paris).

Schwarz, K. H. A.

Papers. Gesammelte mathematische Abhandlungen. (2 vols., ed. K. Schwarz.) 1890, Berlin: reprinted Cleveland.

1872a. "Zur Integration der partiel Differentialgleichung $\frac{\partial^2 u}{\partial x^2}+\frac{\partial^2 u}{\partial y^2}=0.$" *Journ. rei. ang. Math.*, 74(1872), 218–253; *Papers*, 2, 175–210.

Sebestik, J.

1964a. "Bernard Bolzano et son mémoire sur le théorème fondemental de l'analyse." *Rev. hist. sci. appl.*, 17(1964), 129–135.

Seidel, P. L.

1848a. "Note über eine Eigenschaft der Reihen, welche discontinuierliche Funktionen darstellen." *Abh. Akad. Wiss. Munich*, 7(1847–49), math.-phys. Kl., 381–393; Ostwald's Klassiker, 116 (ed. H. Liebmann. 1900, Leipzig), 35–45.

Seidlerová, I.

1961a. "Bemerkung zu den Umgängen zwischen B. Bolzano und A. Cauchy." *Čas pěst. mat. fys.*, 87(1962), 225–226.

Smith, H. J. S.

1875a. "On the integration of discontinuous functions." *Proc. Lond. Math. Soc.*, (1)6(1875), 140–153; *Collected Mathematical Papers*, 2(1894, Oxford), 86–100.

Stäckel, P. G.

1900a. "Integration durch imaginäres Gebiet. Ein Beitrag zur Geschichte der Funktionentheorie." *Bibl. Math.*, (3)1(1900), 109–128.

1901a. "Beiträge zur Geschichte der Funktionentheorie in achtzehnten Jahrhundert." *Bibl. math.*, (3)2(1901), 111–121.

1901b. "Dirichlet'sche Integral." *Ber. Verh. Gesell. Wiss. Leipzig, math.-phys. Kl.*, 53(1901), 147–151. French trans. in *Nouv. ann. math.*, (4)2(1902), 57–63.

Stokes, G. G.

1847a. "On the critical values of the sums of periodic series." *Trans. Camb. Phil. Soc.*, 8(1849), 533–583; *Mathematical and Physical Papers*, 1(1880, Cambridge), 236–313.

Stolz, O.

1881a. "B. Bolzano's Bedeutung in der Geschichte der Infinitesimalrechnung." *Math. Ann.*, 18(1881), 255–279.

1883a. "Zur Geometrie der Alten, inbesondere über ein Axiom des Archimedes." *Math. Ann.*, 22(1883), 504–519.

Struik, R. and Struik, D. J.

1928a. "Cauchy and Bolzano in Prague." *Isis*, 11(1928), 364–366.

Sylow, L.

1902a. "Les études d'Abel et ses découvertes." Abel *1902a*, 1–59 [separate pagination].

Tannery, J.

1908–09a. [Ed.] "Correspondance entre Liouville et Dirichlet." *Bull. sci. math.*, (2)32, part 1(1908), 47–62 and 88–95; and (2)33, part 1(1909), 47–64.

Taton, R.

1947a. "Abel et l'Académie des Sciences." *Rev. hist. sci. appl.*, 1(1947), 256–258.

1947b. "Une correspondance mathématique inédite de Monge." *La rev. scient.*, (1947), 963–989.

1950a. "Un mémoire inédit de Monge: Réflexions sur les équations aux différences partielles." *Osiris*, 9(1950), 44–61.

1951a. *L'oeuvre scientifique de Monge.* 1951, Paris.

1953a. "Sylvestre-François Lacroix (1765–1843): mathématicien, professeur et historien des sciences." *Act. Cong. Int. Hist. Sci.*, 7(1953), 588–593.

Tauber, A.

1897a. "Ein Satz aus der Theorie der unendlichen Reihen." *Monatsh. Math. Phys.*, 8(1897), 273–277.

Taylor, B.

1713a. "De inventione centri oscillationis." *Phil. Trans. London*, 28(1713), 11–21.

1713b. "De motu nervi tensi." *Phil. Trans. London*, 28(1713), 26–32.

1715a. *Methodus incrementorum directa et inversa.* 1715, London; 1717, London.

Terrazini, A.

1957a. "Cauchy a Torino." *Rend. Sem. Mat. Univ. Polit. Torino*, 16(1956–57), 159–203; and 17(1957–58), 81–82.

Tetmayer, J.

1851a. *Théorème général sur la convergence des séries.* 1851, Paris.

Thwaites, B.

1960a. [Ed.] *Incompressible Aerodynamics.* 1960, Oxford.

1967a. "1984: mathematics ⇔ computers?" *Bull. Inst. Math. Applics.*, 3(1967), 134–160.

Timchenko, I. (Тимченко, И.)

1892–99a. "Основанія теоріи аналитическихъ функцій." *Зап. мат. отде. новоросс. общ. естест.*, 12(1892), 1–256; 16(1899), 257–472; and 19(1899), i–xv and 473–655. [Journal sometimes catalogued as "Mémoires de la Société des Naturalistes de la nouvelle Russie."] Reprinted: 1899, Odessa.

Truesdell, C. A.

1960a. The Rational Mechanics of Flexible or Elastic Bodies 1638–1788. Euler *Works*, (2)11, sect. 2(1960).

Valson, C. A.

1868a. La vie et les travaux du baron Cauchy. (2 vols.) 1868, Paris.

Weierstrass, K. T. W.

Works. Mathematische Werke. (7 vols., ed. K. Weierstrass and others.) 1894–1915, Berlin: reprinted 1967, Hildesheim and New York.

1841a. "Zur Theorie der Potenzreihen." 1841, manuscript; *Works*, 1, 67–74.

1856a. "Über die Theorie der analytischen Facultäten." *Journ. rei. ang. Math.*, 51(1856), 1–60; *Works*, 1, 153–221.

1872a. "Über continuierliche Functionen eines reellen Arguments, die für keinen Werth des letzteren einen bestimmten Differentialquotienten besitzen." 1872, manuscript; *Works*, 2, 71–74.

1880a. "Zur Functionlehre." *Monatsb. Akad. Wiss. Berlin, math.-phys Kl.*, (1880), 719–743; *Works*, 2, 201–233.

Winter, E. J.

1949a. Leben und geistige Entwicklung des Sozial-ethikers und Mathematikers Bernard Bolzano (1781–1848). 1949, Halle.

1956a. "Der böhmische Vormärz in Briefen B. Bolzanos an F. Přihonský (1824–1848)." *Veröff. Inst. Slav., Dtsch. Akad. Wiss. Berlin*, 11(1956).

Woodhouse, R.

1803a. The Principles of Analytical Calculation. 1803, Cambridge.

Wussing, H.

1964a. "Bernard Bolzano und die Grundlegung der Infinitesimalrechnung." *Zeitschr. Gesch. Naturwiss. Techn. Med.*, (1964), 57–73.

Yushkevich, A. P. (Юшкевич, А. П.)

1947a. "О возникновении понятия об определенном интеграле Коши." *Труды инст. ист. естест.*, 1(1947), 373–411.

1959a. "Euler und Lagrange über die Grundlagen der Analysis." *Sammelband der zu Ehren des 250. Geburtstages Leonhard Eulers* (1959, Berlin), 224–244.

NAME INDEX

Names preceded by an asterisk are cited in entries or subentries of the subject index.

SUBJECT INDEX

Items are listed as far as possible under main headings or titles: thus there are many subentries for "convergence," "convergence tests," Fourier series," "functions," "integral," and "series." But particular theorems are cited as subentries under the name of the person after whom they are known.